AF701694

Canadian Railway Telegraph History

By Robert Burnet

Canadian Cataloguing in Publication Data

Burnet, Robert G. (Robert George)
Canadian Railway Telegraph History

Includes bibliographical references and index.
ISBN 0-9680243-0-0

1. Railroads - Canada - Telegraph - History.
2. Telegraph - Canada - History. I. Title.

TF627.B87 1996 384.1'0971 C96-930249-5

Second Edition and printing - December 1997
Copyright 1996 Robert George Burnet
ALL RIGHTS RESERVED.

Without limiting the rights under Copyright reserved above, no part of this publication may be reproduced in any form, stored in or introduced into a retrieval system, or transmitted in any form or by any means (electronic, mechanical, photocopying, recording or otherwise), without the prior written permission of the copyright owner.

Care has been taken to trace ownership of previously copyrighted material in this book. The author will gladly receive any information which will enable him to correct any reference or credit line in subsequent editions.

Typeset, Printing and Cover Design by

QUALITY PLUS PRINTING & GRAPHICS LTD.
Mississauga, Ontario, Canada

Telegraph Key & Sounder
P.O. Box 40526
5230 Dundas Street West
Etobicoke, Ontario, Canada
M9B 6K8

TABLE OF CONTENTS

INTRODUCTION

For far too long, the history of the Canadian Railway Telegraph has been neglected. A project of this scope demands a considerable amount of information. Since this aspect of our Canadian Railways has not been studied to any extent, sources are scattered, vague and information conflicting. The search for accurate details and materials has tried to be inclusive. Often only one source could be found. Regrettably, many resources are lost. As a result, personal reflections of those men and women with first hand experience using the telegraph have added considerable authenticity in ambiguous areas. To provide a rounded examination of telegraphic use in Canada, American origins and technological developments needed examination. The reader is encouraged to explore these revealing technological aspects in more depth.

As with all technology, improvements are on going. New and faster ways to get information out are sought resulting in the supplantation of pioneering methodology. The end of the telegraph then, is not an old story, just the same story in a different situation. However, it is long overdue for The Telegraph to take its place both historically and socially in Canada's rich, exciting and evolving railway history.

Space and costs have put restraints on some areas. Therefore, only mainline railway circuits across Canada were examined. Literally hundreds of independently run local telegraph companies existed for commercial and business interests and were quite separate from the railway network.

ACKNOWLEDGEMENTS

Several people have made this project possible through their support, insights, and assistance with research materials - it is very much appreciated with infinite thanks going to: Mrs Connie Romani and her Staff at Canadian National Railways Public Affairs and Advertising; Stephen Lyons at Canadian Pacific Archives; The National Archives of Canada; the Ontario Archives (Toronto); George Copeland, Angela Herwin, George Horner, Gloria Sutherland; Art Clowes of Rail and Transit Magazine (Upper Canada Railway Society), and especially Paul O'Brien, Bev Harvey and John Hann.

Any errors of fact or interruption are the author's responsibility. There is always room for improvement and further development. The search will not end with this volume. Constructive comments and positive criticisms are appreciated.

Etobicoke, Ontario - April 24, 1996

DEDICATION

To my railway family who instilled a deep love and abiding faith in our railways beginning at Finch Station, Ontario, 1845

To Holly, and
To the few people who know and understand my eccentricities

Canadian Railway Telegraph History

CHAPTER ONE

THE DEVELOPMENT AND FORMATION OF THE CANADIAN NATIONAL RAILWAYS TELEGRAPH COMPANY 1846 to 1956

I.

BACKGROUND

British colonists immigrating to Canada in the 1700s had minimal known territory to settle. To the north, the only entrance so far discovered had been through Hudson's Bay. To the south via the Mississippi River and its tributaries, man could penetrate into Hudson Bay Company land. In the east, the St. Lawrence River system provided an entrance to the vast network of the Great Lakes and rivers. As more pioneers arrived, fanning out along the major waterways and river systems, they kept long distances between settlements. On the west coast, the Rockies and the Selkirk Mountains halted any adventurous attempts eastward. Most exploration concentrated on a westward passage to the Orient.

Initially, ships, boats, canoes and canals were the primary means for transportation and communication between independent communities. In 1830, two main water trade routes existed. The first went down the Richelieu River, through Lake Champlain and into the Hudson River. The other, where there was a higher population, was along the St. Lawrence River into Lake Ontario. It followed that the settlers would progress along these easily accessible waterways. For travel or business needs, the steam boats on the St. Lawrence and Great Lakes took travellers almost to where they needed to go. Water travel was inexpensive, faster, and more comfortable than travel on mud packed trails and paths.

The first constructed roads in the Canadas came after the British conquest of 1713. Roadways, initially carved into the land, paid little heed to planned surveys. Before 1800, roads were poor and inadequate. Canada was still a vast frontier land where communication between scattered communities was not a necessity. Early roads, first used for mail, the occasional traveller, and light freight goods over a long distance, "... would [eventually] benefit both agricultural communities and the urban centres they supplied."[1] Locally, there was one main road through town leading to the surrounding rural area.

There were a few major stagecoach routes in the early 1800s. Between Montreal and Kingston, Barnabas Dickson began operating the first long distance stagecoach operation in 1816. The best travelling time on this route took about three days to complete. Service between Kingston and York began in 1817. This stagecoach route, operated by Samuel Purdy, took three to four days to complete.

Stagecoach travel was seasonal, the best time during winter. Since the ground was frozen, a sleigh made good time between settlements. A stagecoach called "The True Briton" had weekly sleigh service between Halifax, Nova Scotia and Windsor, Ontario in 1821. There were many other winter sleigh routes profiting during the winter season. However, when the snow began melting in spring, the pathways turned impossible. The 1820 Emigrant's Guide to Canada states generally:

> The roads are few and poor, but they are moderately commensurate with the retarded progress of the province. The Conveyances, where there are any, are generally poor, the surface rough, the bridges wretched, and the attendance at the inns as defective amidst a scattered, independent, and uncultivated people.[2]

By 1847 it still took at least three days by stagecoach to travel between Kingston and Toronto (York). Stagecoach travel had not improved by 1862, except more routes were available.

FIGURE ONE: Fundamental Mail and Stage Coach Routes 1862

Eastward Destination	Time	Leaves
Kingston	daily, 6 P.M.	Front and Wellington
Oshawa	daily, 3 P.M	Front and Wellington
Rouge	daily, 4 P.M.	Stroud's Market Square (except Sunday)
Westward Destination	**Time**	**Leaves**
Hamilton	daily, 6 P.M.	Hamilton & Lake Simcoe Stage Office along Dundas Street
	Winter 9 A.M.	Hamilton & Lake Simcoe Stage Office along Lake Shore Road
Streetsville	daily, 3 P.M. (except Sunday)	Kellogg's Colbourne Street
Northern Destination	**Time**	**Leaves**
Holland Landing	7 A.M. (except Sunday)	Liddell's Building, Church Street (connection to Lake Simcoe steamer)
Pine Grove	3 P.M.	Liddell's Building, Church Street

(City of Toronto Directory, 1862-1863)

Before 1800, no national consciousness existed. The primary link was a common loyalty or tolerance of the English Crown and a constantly watchful eye on the annexation attempts of the United States. In the Maritimes, trade went to thirteen American colonies. People were content to stay independent and only travel to distant towns if the need was essential. However, in these towns and cities, a slow but constant desire to expand trading opportunities was developing. Canadian businesses were looking for new and closer commercial partners to home than far off England and Europe.

II.

THE TORONTO, HAMILTON, NIAGARA and ST. CATHARINES ELECTRO-MAGNETIC COMPANY

Two revolutionary inventions would eventually help unite the British colonies into one nation. The first was the coming of the Champlain and St. Lawrence Rail Road, running between Lapraire on the south bank of the St. Lawrence River opposite Montreal to St. John's on the Richelieu. This fourteen-mile rail line opened on 21 July 1836. It is noteworthy that by definition of the time, Canada's first railway was a commercial but seasonal portage railway that initially closed each winter. The second revolutionary invention occurred May 1844 when Samuel Morse's electromagnet telegraph became an accepted utility. Its importance for business and trade was realized quickly in the United States. Canada too, recognized "... the importance of a quick connection between all parts of the country as well as the United States."[3]

In 1793, John Graves Simcoe founded York (Toronto). By 6 March 1834, York had become an incorporated city with a growing population of about 9,254 people. Business near the Lake Ontario waterfront was flourishing and expanding west and north. Realizing the potential markets along Lake Ontario, Toronto wholesale hardware merchant, Thomas D. Harris, and Clarke Gamble, a Toronto barrister, turned the sod for Canada's first telegraph company in October 1846: The Toronto, Hamilton, Niagara, and St. Catharines Electro-Magnetic Telegraph Company (THN & St.C). With Capital of 4,000 pounds, 40 pounds sterling a share, and a Charter from the Crown, they financed and constructed their own poles and wire.

The actual construction of the THN & St.C. telegraph line was contracted to Samuel Porter, an American telegrapher who had mastered Morse's code and the construction methods of Ezra Cornell. The specific route is not known since maps were not drawn of telegraph lines until sixty years later. From limited Globe and Mail newspaper accounts, the first telegraph line followed the Lake Shore corduroy and mud-packed road from Toronto through the towns and villages of Port Credit, Oakville, Bronte, Burlington, Hamilton and St. Catharines. It is recorded that two circuits were constructed from St. Catharines: one circuit went from St. Catharines to Fort George and Queenston, from here was "... suspended a wire across the Niagara River to connect the wire at Queenston with Lewiston, N.Y. ... [and it was] ... an excellent, serviceable line."[4] The second circuit went from St. Catharines to Fort Erie then Buffalo, New York.

Samuel Porter operated the first Canadian Telegraph Office in Toronto for the THN & St. C. Company. The head office was at 22 Front Street opposite Browne's Wharf in Toronto. Porter operated approximately 89 miles of single circuit wire. He served as manager, Morse operator, clerk, accountant and messenger boy. Percy B. Marling, also from the United States, was the Company's first telegraph Superintendent, a large shareholder and trained Morse operator.

Between 1846 and 1851, the THN & St. C. Company extended its circuit from Toronto to Quebec City. However, by July 1852 the Montreal Telegraph Company absorbed the THN & St. C.

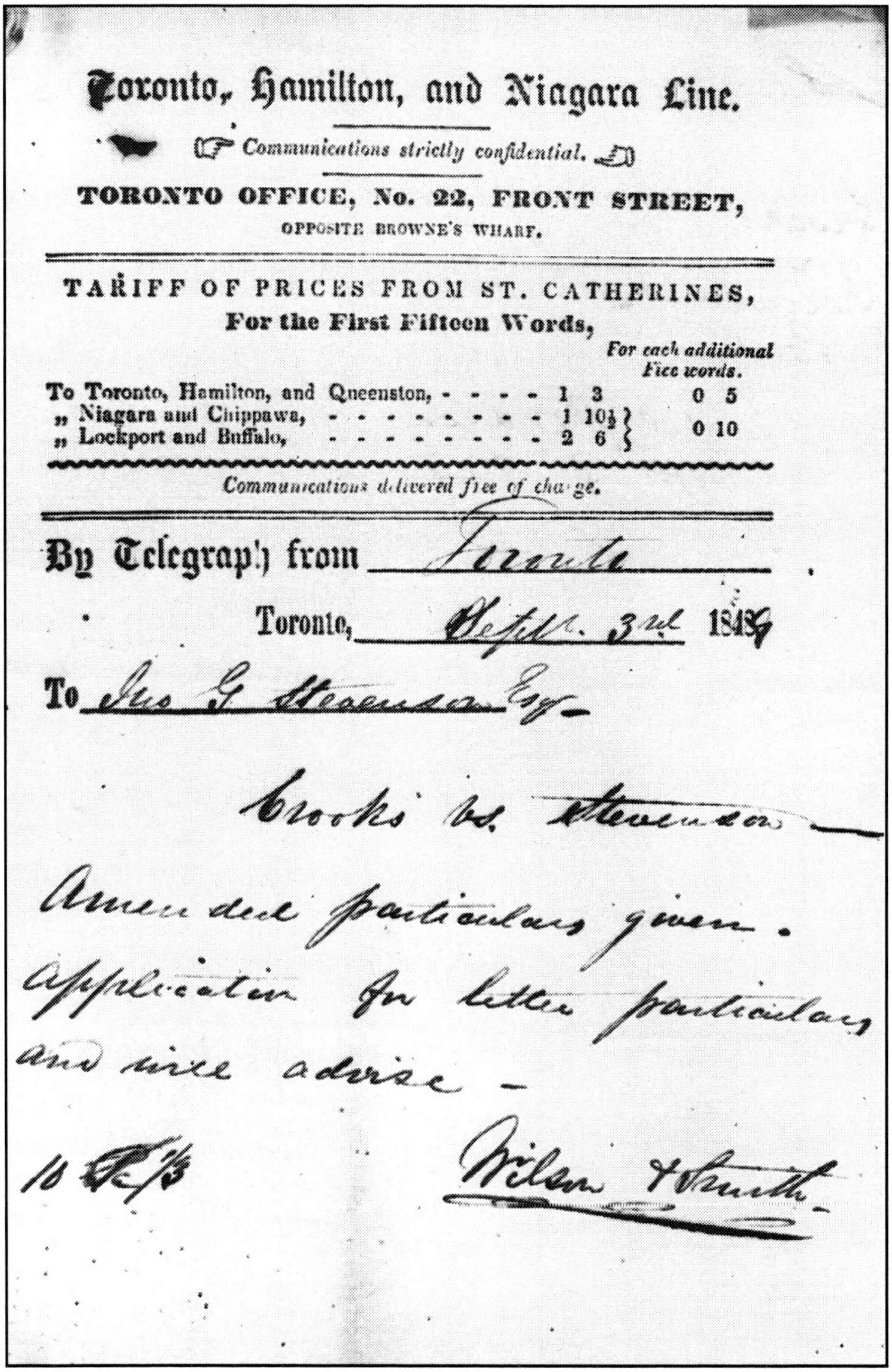

Toronto, Hamilton, and Niagara Line.

Communications strictly confidential.

TORONTO OFFICE, No. 22, FRONT STREET,

OPPOSITE BROWNE'S WHARF.

TARIFF OF PRICES FROM ST. CATHERINES,

For the First Fifteen Words,

	For the First Fifteen Words	For each additional Five words.
To Toronto, Hamilton, and Queenston,	1 3	0 5
„ Niagara and Chippawa,	1 10½	0 10
„ Lockport and Buffalo,	2 6	

Communications delivered free of charge.

By Telegraph from Toronto

Toronto, Sept. 3rd 1849

To Jno G. Stevenson Esq.

Crooks vs. Stevenson —

Amended particulars given.

Application for better particulars and will advise —

10 [illegible]

Wilson & Smith

The Toronto, Hamilton, Niagara and St. Catharines Electro-Magnetic Company - Sept. 3, 1849 Telegram. Note spelling error in the headmast for St. Catharines - correct spelling is with an "a". Note the tariffs in pounds sterling and the Toronto office address. The city was named after Mrs. Catharine Hamilton by her husband, Honourable Robert Hamilton Catharines. He was the Superintendent of the Western District of the Butler's Rangers during the Revolutionary War.

(National Archives of Canada, CNR Collection, PA 189990)

III.
THE MONTREAL TELEGRAPH COMPANY
THE DOMINION TELEGRAPH COMPANY
- and -
THE GREAT NORTH WESTERN TELEGRAPH COMPANY OF CANADA

The Council of the Board of Trade in Montreal met on 26 December 1846. Their objective was to organize an electromagnetic company similar to the Toronto, Hamilton, Niagara and St. Catharines Company. The line would be constructed from Montreal to Toronto. On 29 December 1846, it was determined that sufficient Capital of "... 12,000 [pounds sterling or $60,000] in 1,250 shares of 10 [pounds] each."[5] Further, it was suggested the company be organized under the herald of the Montreal and Toronto Magnetic Telegraph Company. It is noteworthy that "... a liberal charter by the Quebec Provincial Legislature ... [was not obtained until] ... July 28, 1847..." [6] giving the telegraph company powers to acquire smaller telegraph companies as well as extending their circuits throughout Quebec. Also, under the charter, the original company name (Montreal and Toronto Magnetic Telegraph Company) was changed to the Montreal Telegraph Company. Interestingly, the Montreal Telegraph Company became the "... earliest electric corporation in Montreal ... but which, as a mere line of passage, scarcely affected the city [at first]." [7]

With charter and capital, the Montreal Telegraph Company began construction at the end of July 1847. Orrin S. Woods, the first Superintendent of the company was in charge of operations. Woods, appointed 6 March 1847, remained in the Superintendent position until he retired in 1865. Woods, who had been Samuel Morse's first telegraph student, hired the contractor Livingston and Wells, an express-forwarder based in New York, to construct the first line west from Montreal. Simultaneously, a line went east to Quebec City. Woods purchased the best materials for the construction of the telegraph line. The poles were

> ... almost solely of large cedar, planted not less than five feet in the ground, were thoroughly tamped and ten rods apart [about 17.5 feet]. Wooden brackets of seasoned white oak with glass insulators ... were affixed to the poles an either side, and on these were mounted two English [iron] galvanized wires, number nine, the first of the kind employed on the continent ....[8]

Orrin Woods demanded the finest available operating telegraph equipment. The manufacturing company of A.S. Chubbuck and Son of Utica, New York (ca1847-1869, located on Hotel Street), made and supplied Morse equipment to many American telegraph companies, such as Western Union. Woods had designed a telegraph model that Chubbuck produced. These models were "... of the finest workmanship, strong and durable."[9] The receiving equipment was the recording paper tape register device, the precursor to the sounder.

From Quebec City to Montreal and west to Toronto, the poles of the MTCo went through the major centres of Kingston, Belleville, Toronto, Hamilton, St. Catharines, Niagara and Buffalo, in addition to smaller towns and villages in between. The Montreal Telegraph established telegraph offices not only in major towns and cities, but also targeted the small towns, villages and hamlets. Interestingly, the THN & St. C. had only established telegraph offices in the larger centres.

The Montreal-Toronto circuit was completed 3 August 1847, and "... it was found to be satisfactory in every particular, both in material and workmanship, and became a model for future work of the kind."[10] The Montreal-Quebec City circuit was finished 2 October 1847.

By December 1847, the Montreal Telegraph Company had constructed 540 miles, opened nine offices, logged approximately 33,000 transmitted messages, and had thirty-five employees. The telegram rate between Toronto and Quebec City was one dollar for a ten-word message.

In 1851, the Provincial Legislature was preparing a new law that would have significant effect on establishing a telegraph company. A new telegraph company under The General Law for Electric Telegraph Companies, received specific legal rights. Government charter's were required for any type of construction involving Crown lands and to obtain financial support for small financial assistance or guarantees. When the Act passed in 1852, it established a political and corporate tone: "An Act to provide by one General Law for the Incorporation of Electric Telegraph Companies. Whereas it is expedient to provide one General Law for the Incorporation and regulation of Companies formed for the purpose of constructing lines of Electric Telegraph in this Province:...."[11] The Act then began a point-by-point legal outline (see Appendix Two for the complete Act).

The Electric Telegraph Act required no less than three people associated with the company to organize the telegraph circuit and any branches leading to or from the originating office, but only upon the terms and conditions detailed in the Act. Other considerations demanded the company make and provide a sealed certificate specifying the Company name, route or routes; capital stock, shares and shareholder names and amounts; the period of construction from commencement to termination; a copy of the company By-laws, Articles and provide the minutes of meetings on demand.

The Crown provided clauses in the Act giving the company specific legal rights. Some of the more significant included the "... power to purchase, reserve and hold and convey, such real estate and such only as may be necessary for the convenient transaction of the business and for effectually carrying on the operations of such Association...."[12] The Crown would provide protection to a company against vandalism to facilities. They would also prosecute those caught.

A liberal authorizing clause concerning routes was available in the 1852 Act:

> ... to construct the lines of Telegraph designated in its Certificate, along any and upon any of the public roads and highways, or across any of the waters within this Province [Ontario], By the erection of the necessary fixtures, including posts, piers or abutments for sustaining the cords or wires of such lines, provided the same shall not be so constructed as to incommode the public use of such roads or highways, or injuriously to interrupt the navigation of such waters....[13]

Early telegraphic companies were regarded with scepticism. Since the idea of telegraphy was new, it was assumed most companies would not become profitable and larger companies would take over the smaller ones when the opportunity was right. The law therefore included a proviso that it would "... not be lawful ... to contract debts exceeding one half of the amount of the Capital Stock ...."[14]

A later clause protecting the Crown's interests and investments stated: "Her Majesty may, at any time after the commencement of any telegraph line under this Act, and after two month's notice to the Company, assume the possession and property thereof ... [and] ... shall be vested in the Crown."[15]

Finally, the Act added an additional guarantee for the government in case of war: "Her Majesty may at any time, upon giving one month's previous notice to the Company, assume possession of any such Telegraph line and of all things necessary to the sufficient working thereof, for any time to be mentioned in such notice, and may for the same time require the exclusive service of the Operators and other persons employed in working such line ...."[16] This clause, later amended, was invoked twice during the next few decades: when the Fenians stormed across the border into Fort Erie, and Louis Riel's Prairie rebellions.

Meanwhile, the Montreal Telegraph Company, between 1850 and 1852, established four main objectives. The Company had been operating for three years now. Andrew Shaw, appointed the first president on 28 July 1847, achieved much for the original organization of the company. But the shareholders were demanding more dividends. By 1850, the primary objective of the Board of Directors was the election of an aggressive new president. In 1851, Hugh Allan was first nominated then voted into the office of president, a role he held without challenge until his death. Allan (B: 29 September 1810 / D: 9 December 1882) secured the skills of O.S. Woods, who became his right-hand-man and the original Secretary, James Dakers, who would later be put in charge of the Eastern Region.

The second planned objective was westward expansion to British Columbia. About 1850, H.P. Dwight, originally a telegraph operator in Montreal, was placed in charge of the Toronto office by Allan. Allan wanted Dwight to watch for and obtain as needed any possible western expansion into the Hudson Bay Company's land. "With 'carte blanche' to make whatever extensions he might deem necessary or advisable, Mr. Dwight proceeded to cover the Province [Ontario] with a network of wires, built necessarily on the public highways, no railway routes being available for this purpose at that early [time]."[17] Dwight managed the Toronto office with the help of only one office-messenger boy, R.H. Hathaway. Hathaway later wrote that Dwight sent, received all messages numbering twenty to thirty a day, and even delivered them.[18] For almost sixty years, Hathaway worked for Dwight. In 1865, O.S. Woods retired, and Dwight was placed in charge of the Western Region of the Montreal Telegraph Company. It is interesting that Dwight did not see any economical or justifiable benefits in constructing a telegraph circuit westward through northern Ontario, the Prairies to British Columbia. Dwight argued that since the population in Hudson Bay land was widely scattered and most people were Indians, Metis and trappers, telegraphic use would be rare. Therefore, the second objective was altered by Dwight to concentrate solely on the mainline circuit from Quebec to Sarnia and Windsor.

Allan's third goal would be the development of circuit extensions off the main trunk line built along the St. Lawrence River and Lake Ontario waterways into remote regions of the provinces. O.S. Woods stated that branches were directed towards "... the most remote towns and villages placed in telegraphic connection with the chief business centres of the country."[19] Further, "Whenever a little hamlet wanted to be connected with the exterior world, or a distant port wished to be within electrical hail of an interior market, its cry was readily heard, and the wished for line soon appeared."[20] The Company knew remote lines would be unprofitable, but they reasoned the main line profits would offset any loss. This initiative would politically and economically unite Upper and Lower Canada before the railway's influence.

The fourth telegraph objective involved the railway. In 1851, the question of the main route of the Grand Trunk Railway was debated. The Standing Committee of the Canadian Railroads and Telegraph Lines had many requests for the main line tracks that ultimately would effect the telegraph. Two primary routes debated were: Montreal-Ottawa-Toronto and Montreal-Kingston-Toronto. After considerable contrariety of opinion, especially by President Hugh Allan of the Montreal Telegraph Company who demanded the Montreal-Ottawa-Toronto railway route, the ultimate railway right-of-way was determined by the Standing Committee.[21] The members concluded that a "...short line was the best for the direct trade; and this outweighed everything else in the selection of the route [between Montreal and Toronto through Kingston]."[22] Interestingly, the telegraph already existed between Montreal-Kingston-Toronto. It cannot be satisfactorily explained why Allan would insist on a route from Montreal-Ottawa-Toronto for the railway. Next, in 1852, the GTR was given a charter to build their tracks between Montreal and Toronto along with rights to construct a telegraph line parallelling the tracks. Hugh Allan's Montreal Telegraph Company was selected by the Grand Trunk to construct the railway telegraph system.

The original Montreal Telegraph route penetrated inland parallelling the Macadamized and corduroy roads. Not until the Grand Trunk Railway began building west in 1852, did telegraph poles and wires begin to find themselves constructed beside Canadian railway right-of-ways. For the telegraph company this was an exceptional arrangement as the railway contracted surveys and logical routes for their tracks. Telegraph companies then bargained for the right to have the railway put in the poles and charge the telegraph company or the telegraph company would put in the poles and charge the railway for a circuit. Either way, the telegraph company cut costs and avoided expensive land appropriation by following railway right-of-ways: thus began the fraternity of telegraph poles beside the railway line. Also, repair to the poles and wires was easier along a railway right-of-way than sending out an operator by horse over a separate or similar route and who may not be seen for a week at the station. It is significant to note that the Grand Trunk Railway survey plans between Montreal and Toronto would parallel the established route of the MTC poles along the Lake Shore waterway. Through this arrangement with the railway though, many telegraph companies would eventually lose their identity.

During 1851, the shareholders of the Toronto, Hamilton, Niagara and St. Catharines Electro-Magnetic Company, began voicing strong disapproval of their Company's low dividend payments. The shareholders, furthermore, became concerned about the Montreal company and its circuit along the St. Lawrence-Lake Ontario lake shore and the Grand Trunk Railway proposals. The Montreal Telegraph displayed forcefulness in all their demands, and financially they were much stronger, giving them an obvious business advantage. As a result, "... some subscribers [of the Toronto, Hamilton, Niagara and St. Catharines Electro-Magnetic Company, were] dissatisfied with the conduct of business, attacked the management [of the Montreal Telegraph Co.] through newspapers, but the columns of controversy failed to halt the progress of work [by the MTCo]."[23] However, the THN & St. C. ploy faltered and failed and can be confirmed on 29 July 1852, by the Huron Signal in Goderich, Ontario: "... the recent consolidation of the Toronto Hamilton Buffalo Telegraph Co. [sic] with the Montreal Telegraph Company ..."[24] had been accomplished. Canada's first telegraph company was now absorbed by the Montreal Telegraph.

Montreal Telegraph Company,

OFFICE,

22 Front Street,

CONNECTING WITH ALL THE PRINCIPAL

CITIES AND TOWNS IN

CANADA AND THE UNITED STATES.

All Communications strictly confidential.

Write plain— Give full Address of yourself and correspondent—Answer promptly by this Line to avoid delay and confusion.

Company not responsible for the inaccuracies of the Operators.

Toronto, October 24th 1852.

By Telegraph from Hamilton, Ontario

To George Barrow, Toronto, Ontario

Buy shares at prices quoted in your telegram of date and confirm

J. Johnson

The above message blank is a facsimile of that in use by the Montreal Telegraph Co., a century ago. To commemorate the centenary of **THE TORONTO STOCK EXCHANGE**, the Canadian National Telegraphs, successor to the Montreal Telegraph Co., reproduces this pioneer telegraph form on which quotations and messages were written when **THE TORONTO STOCK EXCHANGE** was established, in 1852.

Montreal Telegraph Company Telegram, 24 October 1852. Note the headmast address confirming the Montreal Telegraph Company takeover of the Toronto, Hamilton, Niagara, St. Catharines and Niagara Company. The Montreal Telegraph continued to use the same office site. (CNR Public Affairs, #48119)

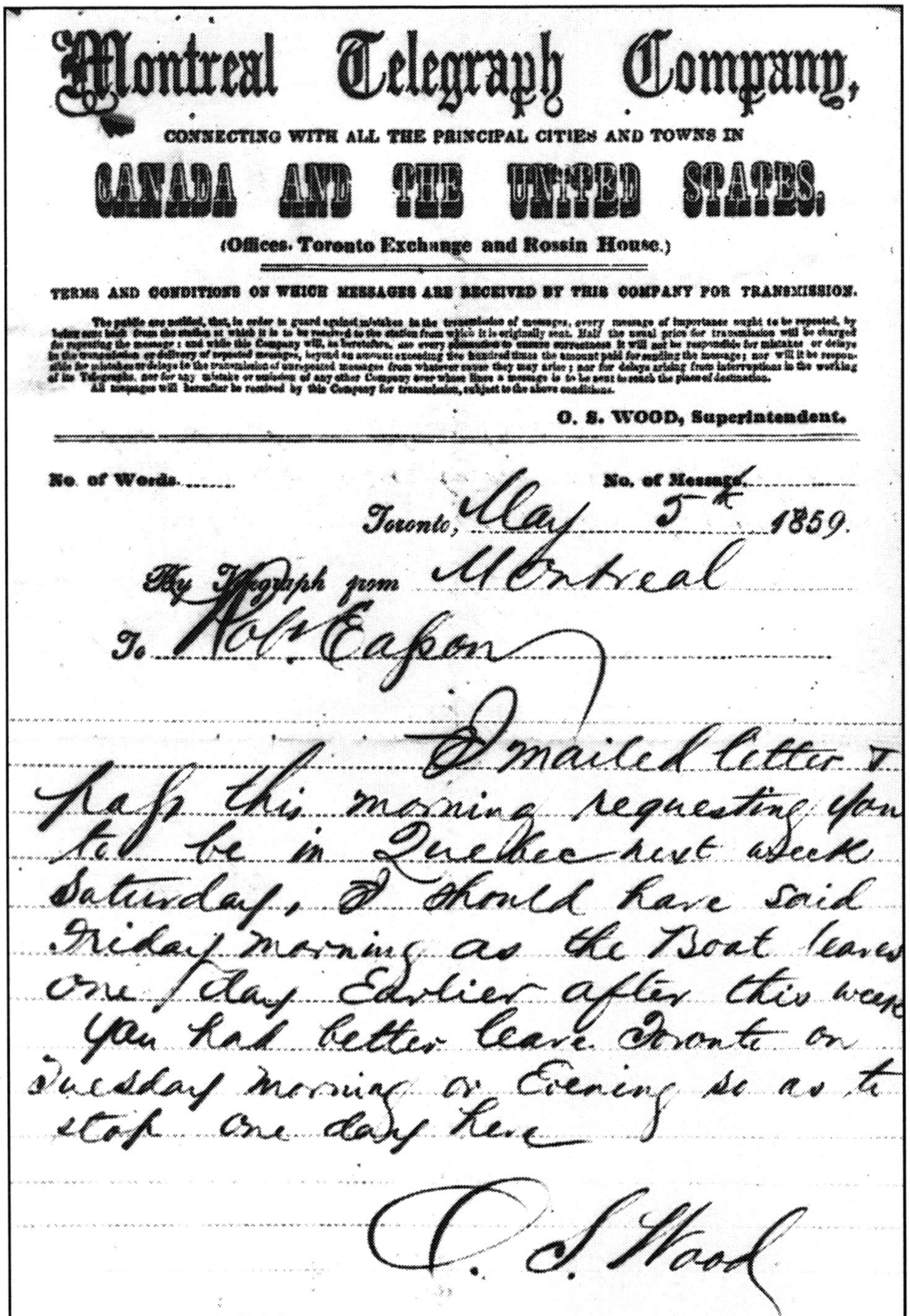

Montreal Telegraph Company,

CONNECTING WITH ALL THE PRINCIPAL CITIES AND TOWNS IN

CANADA AND THE UNITED STATES.

(Offices. Toronto Exchange and Rossin House.)

TERMS AND CONDITIONS ON WHICH MESSAGES ARE RECEIVED BY THIS COMPANY FOR TRANSMISSION.

The public are notified, that, in order to guard against mistakes in the transmission of messages, every message of importance ought to be repeated, by being sent back from the station at which it is to be received to the station from which it is originally sent. Half the usual price for transmission will be charged for repeating the message; and while this Company will, as heretofore, use every precaution to ensure correctness, it will not be responsible for mistakes or delays in the transmission or delivery of repeated messages, beyond an amount exceeding five hundred times the amount paid for sending the message; nor will it be responsible for mistakes or delays in the transmission of unrepeated messages from whatever cause they may arise; nor for delays arising from interruptions in the working of its Telegraphs, nor for any mistake or omission of any other Company over whose lines a message is to be sent to reach the place of destination.
All messages will hereafter be received by this Company for transmission, subject to the above conditions.

O. S. WOOD, Superintendent.

No. of Words............ No. of Message..............

Toronto, May 5th 1859.

By Telegraph from Montreal

To Robt Eaton

I mailed letter & Pass this morning requesting you to be in Quebec next week Saturday, I should have said Friday morning as the Boat leaves one day Earlier after this week. You had better leave Toronto on Tuesday morning or Evening so as to stop one day here.

O. S. Wood

Montreal Telegraph Company Telegram, May 5, 1859 - Note the headmast changes although the logo is the same. The company 'protection clause' has been expanded. The message implies a trip of at least two to three days from Toronto to Montreal.
(National Archives of Canada, CNR Collection, PA 189989)

Telegraph business was first concentrated around the shipping season. Principal customers were forwarders, millers, and grain farmers. For example, in 1847, the British North America Telegraph Company (BNATC) was established. It's primary objective was to report the movement of ships and boats along the St. Lawrence River from Quebec City to Father Point to Riviere-du-Loup. However, the telegraph circuit soon developed into transmitting news releases received from Europe and Britain to Canadian and American newspapers. "However, this line was unfortunately a few years ahead of its time, since ocean steamships were not yet running to Quebec or Montreal, and very few of the sailing vessels passing by, bothered to report themselves by telegraph."[25] Interestingly, despatches from the ships were put into a sealed tin box, then tossed into the water. The landline operator using a small boat, retrieved the box then set about keying the messages. Occasionally, dogs reportedly recovered the despatch box, however, it is not known with accuracy if the dogs were ship or land based or if this was an official practice. By 1851, the BNATC circuit was discontinued through a lack of profitable commercial use. Later, the Montreal Telegraph would revive the circuit for a short time with a shipping and press release service. Albeit, by 1866 with the establishment of the Trans-Atlantic Cable, the ex-BNATC circuit was completely abandoned.

With the advent of the railway, seasonal business began to wane. By 1852, the railway was beginning to carry freight and passengers to more places in all types of weather. At first, the telegraph was used to transmit to distant areas, arrange business meetings, secure deals and orders, but did not have any significant railway involvement. So successful was the Montreal Telegraph with its expansion objectives and business profits, it was soon called upon to construct and operate circuits for railway use.[26] On 18 November 1853 the Montreal Telegraph "... contracted to construct a telegraph line of poles and wire and to supply instruments for operating same at prices stated in the Agreement, along the lines of the Grand Trunk Railway Company of Canada ... [and also provide] ... for its own use, on the pole lines without charge, one wire for the entire length of line and one extra wire between Montreal and Toronto."[27] This proved financially lucrative for the Montreal Telegraph; it too initiated for the first time, sections of telegraphic train dispatching along the GTR right-of-way. It is believed at this time the Montreal Telegraph established two telegraphic divisions: the initial pole route along the St. Lawrence and Lake Ontario shores became a strict commercial circuit; the new poles placed beside the Grand Trunk Railway tracks became the operating and signalling railway circuit.

A new telegraph act was prepared and passed by parliament in 1852. The first company to incorporate under the new General Telegraph Act of 1852 was the Canada Grand Trunk Telegraph Company (CGTTC). The CGTTC was organized as a commercial venture on 16 March 1852, though not incorporated until 14 December 1852 by Provincial Statute, 16th Victoria, Chapter 10. Its Capital Stock was "... limited to 25 [pounds] per Mile, excepting over Water-courses where Gutta Percha is used [a tough greyish-black substance, flexible when thin, made from the juice of various Malayan trees, used for insulation], for which part, the Capital Stock is limited to 125 [pounds] per Mile. Shares Five Pounds each. Stockholders have one vote each share held by them."[28] Their wires would extend from the Toronto head office on Front Street to Buffalo and Quebec City. (The Canada Grand Trunk Telegraph had no affiliations at any time with the Grand Trunk Railway or the Montreal Telegraph Company.)

FIGURE TWO: News Clipping

Barrie, Northern Advance
Wednesday, April 19, 1854

CANADA GRAND TRUNK TELEGRAPH COMPANY

Four Hundred and Forty-Five Miles of the Wires of this Company are now complete and in operation.

Thirty-Eight Offices are now open for business. Dispatches received for the following points, and forwarded with the utmost promptness:

Hamilton
Dundas
Galt
Preston
Berlin
Guelph
Georgetown
Norval
Brampton
Streetsville
Lambton
Barrie

Kingston
Bath
Newburg
Napanee
Demorestville
Picton
Wellington
Consecon
Port Trent
Brighton
Colbourne
Newmarket
and Thornhill

Cobourg
Port Hope
Newcastle
Bowmanville
Whitby
Belleville
Port Credit
Brantford
London
Simcoe
Port Dover
Richmond Hill

Office Front Street, a few doors West of Church Street.
W. D. SNOW.
Secretary.

Note: Every Village is destined to have a Telegraph Office. Messrs. Snow & Dwight, have contracts from various Companies to build 2,800 miles of Telegraph lines, within the next year.

(Toronto, July 8, 1853)[29]

The Canada Grand Trunk Telegraph ran into considerable financial troubles almost immediately. In their 1854 Minutes, more time dealt with the payment of travel money and expenditures to Directors. Most meetings however were not held in Toronto but rather Kingston, Picton, Belleville, or Peterborough, as members voted themselves expense and travel allowances. Meetings were also concerned with telegram tariffs. However, little was discussed about routes or expansion of their network. By 1855, the Montreal Telegraph exercised their Charter rights to acquire the floundering CGTTC circuit.

Other company acquisitions classified the Montreal Telegraph Company as a leading company. Such was its financial strength that "Many companies [which] started to compete in the same line of business, but in no time were doomed to failure, as they neither had the skill, the men nor the money to enable them to succeed in opposing their more formidable rival."[30] Accordingly, more companies became part of the Montreal Telegraph as outlined in Figure Three:

FIGURE THREE:

Significant Telegraph Companies acquired by Montreal Telegraph

TELEGRAPH COMPANY	MAJOR CIRCUIT ROUTE	MTC ACQUIRED DATE and NOTES
British North American Electric Telegraph Association est. 1847	Quebec City to Father Point via Rivere du Loup	ca 1851 when Father Point to Rivere du Loup dismantled
Montreal and Troy Telegraph Company est. 1849	Montreal to Whitehall and Troy	between 1851 and 1853
Prescott-Bytown Telegraph Company est. 1850 (also Bytown-Prescott)	Prescott to Bytown (Ottawa)	J. Aumond establish; ca 1853 Allan wants as this was part of original GTR route
Canada Grand Trunk Telegraph Company incorp. 14 Dec. 1852	Quebec City to Montreal, Kingston Toronto, Buffalo	1855 for $11,000.00

American owned or leased companies joining the list between 1860 and 1875 included:

TELEGRAPH COMPANY	MAJOR CIRCUIT ROUTE	MTC ACQUIRED DATE and NOTES
New Brunswick Electric Telegraph Company incorp. 4 Oct. 1847	Calais to Amherst Halifax to Amherst (Nov. 1849) St. John to Calais and Portland, Maine (Jan. 1, 1849)	American Telegraph ca 1860 then Western Union ca 1866
Nova Scotia Electric Telegraph Company completed 9 Nov. 1851	Digby N.S., to Portland, Maine Halifax to Calais	NS government takes over then American Telegraph in 1860 for 10 years - Western Union purchases 1872
Northwestern Telegraph Company completed 20 Nov. 1871	Winnipeg (Fort Gary) to Pembina	connects with USA Postal Telegraph Company
Canada Mutual Telegraph Company and the Mutual Union Telegraph Company	Toronto to Niagara Falls	see *

* both companies worked in conjunction for a short time in Canada and the United States; their origins are vague, but it has been suggested they were holding companies; both companies were absorbed by Western Union ca 1880 then by the Great North Western Telegraph Company ca 1881-1882. The CPR also leased two wires from the GNWTCo on 1 January 1886 between Toronto and Niagara Falls, Ontario.

By 1861, the MTCompany had a telegraphic monopoly throughout Ontario and Quebec. Much of their success being "... its willingness to expand, and not confine itself to too small an area. The latter was something which many of the smaller companies were forced into doing...."[31]

Between 1862 and 1867, Canadian interests centred on forming a nation. The American Civil War provided the Canadian/British Government with many reasons to protect its border. The most significant motivation included fears of an American military invasion, political annexation and the American migration of soldiers and settlers onto the Canadian prairies.

The other catalyst towards a Canadian Confederation were the many Fenian raids into Ontario, Quebec and the Maritimes in 1866. Specifically, on 31 May 1866, the Fenians attacked Fort Erie, Ontario cutting the communication and commercial telegraph circuit of the Lake Huron and Buffalo Railroad. This circuit branched off to connect with military posts and businesses from St. Catharines to Hamilton, Burlington, Oakville and Toronto. When the line was severed, it demonstrated how vulnerable the Dominion was to invasion. The Fenian assault revealed how vital the telegraph network was in reporting disastrous events quickly so fast and effective action could be provided.

Concurrently, the Montreal Telegraph Company expanded their commercial and railway telegraphic development with the Grand Trunk Railway. The MTC began augmenting their circuits more quickly into remote regions, along GTR branchlines, and towards the larger centres of Sarnia and Windsor because of the invasion anxiety between 1862 and 1866. Not only did this telegraphic expansion assist the MTC profit ledgers, it arguably provided security to communities fearful of assault. Allan took advantage of the invasion climate to push his Company's visibility towards communication dominance in most remote, rural and city centres.

Simultaneously in the Maritimes, the newly completed trans-Atlantic cable in 1866, encouraged the MTC to make further telegraphic connections. The Montreal Telegraph saw Canadian communication nationalism from the Maritime Provinces along the St. Lawrence River and Lake Ontario shores to Windsor and Sarnia. However, there were several political obstacles stunting this telegraphic growth, as expanded in Section V. It is unfortunate Allan did not over-ride Dwight's decision to expand the telegraph through northern Ontario and west across the Prairies to British Columbia. The westward telegraphic and railway expansion would later fall to Edward Watkin and the Canadian Pacific Railway.

By 30 June 1866, the New Brunswick government voted on a Canadian Confederation and the construction of the Intercolonial Railway. Between 4 and 24 December 1866 members from Canada, Nova Scotia and New Brunswick met with the British government in London, England to obtain a commitment for confederation and the Intercolonial Railway. On 12 April 1867, the Imperial Canada Railway Load Act was authorized by Britain, providing a construction guarantee of three million pounds sterling for the Intercolonial Railway between Halifax and Quebec.

By 1868 the Montreal Telegraph Company had purchased most of its significant competition. MTC currently operated lines in Quebec, Ontario, and partly in the Maritimes; connecting circuits existed with many United States telegraph companies. Agreements were also underway with expanding telegraphic communication in the Maritimes along the Intercolonial Railway.

On 9 September 1869, the Montreal Telegraph and GTR reached a new agreement. Based on the original agreement of 18 November 1853 arranged by Hugh Allan, and letters dated 18 and 22 April 1862 with the GTR Managing Director, the Grand Trunk "... sold to the Montreal Telegraph Company for $30,000, the sole and exclusive right and privilege of erecting, maintaining and operating, upon, along and over the lines of the [GTR], with the guarantee of quiet and undisturbed possession forever." [32] Further, in October 1869, Allan "... offered to build a telegraph line from

Riviere-du-Loup to Halifax. This offer was accepted on May 14th, [1870] with the proviso that the Montreal Telegraph Company maintain the line by the virtue of its independent circuit to remain their property, the [Intercolonial] Railway telegraph circuit being strung on the same poles."[33] This circuit was an extension of the defunct British North American Telegraph Company.

On 22 September 1870, the MTC offer became a legal agreement between the Intercolonial Railway and Crown. The MTC would build "... along the entire line of the Intercolonial Railway and Branches ...."[34] The Crown, as partner, would contribute $30 per mile and "... can at any time elect to purchase the Telegraph property at a fair valuation ...."[35] One wire was assigned to the Intercolonial and given the right to a second circuit, at cost to the Montreal Telegraph any time the GTR felt a wire was required; this it exercised on 19 September 1881: "... as the single circuit then in use was overcrowded."[36]

On 1 June 1871, MTC announced they were free of debt and attributed it to building their lines where business and the population wanted it. Major terminal offices were found in Quebec City, Montreal, Ottawa and Toronto. Allan felt confident he could create a uniform system of telegram tariffs. The 1847 telegram rate was still in effect until June 1871: $1.00 for ten words. However, on 1 June 1847, the rate was cut to 25 cents for ten words plus one cent more for each additional word. This tariff covered all circuits from Sackville, New Brunswick to Sarnia, Ontario, all lines north and along Georgian Bay, and the routes to Gaspe and Bonaventure. It also included destinations to the United States, from the Montreal Office, to Portland, Maine (300 miles); Boston, Massachusetts (350 miles); New York, New York (450 miles); Oswego, New York (300 miles); Buffalo, New York (450 miles), and, Detroit, Michigan (550 miles). Business was lucrative.

A innovative competing telegraph company was forming. Selah Reeves, under the General Laws of the new Confederation, began organizing the Dominion Telegraph Company (DTC). The DTC was organized through selling subscriptions around the Clifton area, later renamed Niagara Falls, Ontario. With the necessary estimated capital raised between $700,000 and one million dollars, Reeves secured contracts to build a single circuit of about two thousand miles. However, only three hundred miles were constructed under his control. As a result, disgruntled stockholders met in Montreal and cancelled his contract. However, several investors became determined to keep this company alive. On 11 January 1871, a newly reorganized Dominion Telegraph appeared with an elected Board of Directors and officials. On 14 April 1871, by Dominion Charter, Chapter 82, a fresh DTC was incorporated.

By the mid-1870s, the Dominion Telegraph constructed more terminal facilities, made broader connections in eastern Canada, and established links with the United States. After enacting charter clauses to obtain the necessary right-of-way alongside roadways and over water courses, E. Colby, an American based contractor from Oswego, New York, began constructing the principal DTC circuits in 1871. Cities like Ottawa, Cornwall, Kingston, Belleville, Toronto, Hamilton, London, Windsor, Sarnia, Niagara Falls were established first. Ontario branch circuits included the smaller areas like Guelph, Mount Forest, Barrie, Peterborough; Quebec circuits first included, Montreal, Quebec City, Sherbrooke, and Rivere-du-Loup. DTC service also targeted smaller towns and villages not served by MTC.

During 1875, DTC Maritime connections to Ontario and Quebec were established. Again, by enacting specific charter land and water rights, circuits were constructed beside New Brunswick and Nova Scotia roadways and over waterways. (Further discussion about this service is established in Section V.)

Dominion Telegraph affiliations to the United States included the Atlantic and Pacific Telegraph Company, the Vermont International Telegraph Company plus border connecting circuits to Ontario through Detroit, and Buffalo. Europe was part of the DTC hookup through a circuit with the Direct United States Cable Company and the Trans- Atlantic Cable Company at Tor Bay, Nova Scotia. This cable extended into Quebec City and Montreal.

Also in 1875, the Montreal Telegraph Company, now firmly established, maintained their extensive telegraph network with little to no growth into new communities. MTC operated over 20,000 miles of circuits out of fourteen hundred offices, employed over two thousand men and women, and earned over two million dollars in profit. At the end of 1875, at least two million messages were transmitted; the number received is not recorded. Revenue also grew from innovative departments such as the Press Service. However, MTC scrutinized their competitor, the Dominion Telegraph, with more concern.

As early as 1876, Thomas Swinyard, General Manager, Toronto Office of the Dominion Telegraph wrote: "... there is a suggestion that the Dominion and Montreal Telegraph Companies ... divide the telegraph business in Canada equitably between them. [It would appear that] the Dominion Telegraph Co. was established to destroy the Montreal Telegraph Co.'s monopoly. Amalgamation, therefore, difficult to accomplish, and [would be] universally unpopular and, therefore, impolitic."[37] The suggestion was ignored.

During 1878, a financial depression strangled the country. This economic disaster struck the Dominion Telegraph, and Montreal Telegraph ruthlessly. The DTC had now established their Maritime lines, "... and in an evil moment its management decided, as a bid for popularity and business, to reduce rates between the most distant points in Ontario and Quebec from twenty-five cents ... to twenty cents for ten words."[38] The Montreal Telegraph countered, but found the rate too low to sustain profits. Consequently, the rate war "... was fatal to both companies ... and neither could afford to make further extensions, [to] enterprising places or needy districts...."[39] But exceptions existed: for example, on 13 February 1877, J. C. Bailey, C.E., recommended to the MTC that a circuit could be built from Stouffville to Jackson Point, Ontario, following the Lake Simcoe Junction Railway for only $1,000 - construction was denied. However, MTC opened a new circuit on 25 July 1878 from Matane to Fox River, Quebec with nine intermediate telegraph offices. On 10 August 1878, the Crown approved the construction and paid a $15,000 bonus to Quebec authorities. However, profits for both companies continued to erode significantly.

By a sealed agreement dated 12 June 1879, the Dominion Telegraph was leased to the American Union Telegraph Company for 99 years. This indenture was effected on 1 July 1879. It included all the Dominion telegraph lines and facilities throughout Canada and elsewhere for the following:

- an annual rental fee of $52,500, which could be increased to $62,500 if the Dominion Telegraph pooled with a third telegraph company;
- the company would return five-percent of its paid Capital;
- Rates were set to 20 cents for ten words in October 1879, however, the Montreal Telegraph was not aware of the secret agreement reached with the Dominion Telegraph and the American Union. Montreal Telegraph was exchanging business and appeared to have excellent relations with Western Union. The American Union Telegraph Company, formed by Jay Gould, is credited with lowering the telegram rates in the United States, forcing the Dominion Telegraph to act next; Gould then assigned the Dominion Telegraph holdings to the Western Union Telegraph Company for 99 years on 11 July 1881. Later, Western Union would sublet these holdings to the Great North Western Telegraph Company of Canada while retaining the Dominion Telegraph name.[40]

A telegraph company in Winnipeg appeared, which was not in a geographic position to offer serious competition. The Great North Western Telegraph Company of Canada (GNWTCo) was incorporated under Victoria 43, Chapter 666, on 7 May 1880. A Provisional Directorship managed the affairs until a suitable person for the role of President was found. Capital Stock was established at $400,000 and offered shares at $100 each. The initial circuit would be from the Winnipeg head office to Thunder Bay, along Lake Superior and Georgian Bay to Toronto, Ottawa and Montreal. Another circuit was constructed from Winnipeg to the District of Keewatin. To the south, a circuit ran from Winnipeg to connect with the Western Union at St. Paul, Minnesota.

At the 10 June 1881 shareholders meeting of the Great North Western Telegraph, Erastus Wiman (1834-1904), a Canadian by birth who later became an American citizen, was nominated then elected President of the GNWTCo. On 30 August 1881, a new and expanded directorship elected to manage the company's property: "O.S. Woods, of Montreal [was elected] Vice President. H.P. Dwight ... appointed General Manager; Frederick Roper, Secretary ..., and Arthur Cox Treasurer."[41]

Acting quickly, Wiman affected one of his company's charter rights: to obtain a lease or acquire outright another telegraphic company. Wiman's goal was the complete acquisition of the Montreal Telegraph and Dominion Telegraph with the operating monopoly of the Western Union Telegraph Company of the United States. It was well known, as previously noted, the two Canadian companies were in serious financial trouble. The Western Union had already leased the Dominion Telegraph on 1 July 1879, which it obtained by purchasing the American Union Telegraph Company.

Wiman, in 1881, further empowered by the directors of the GNWTCo, negotiated with the Montreal Telegraph, Dominion Telegraph and Western Union. Speaking for the directors therefore, Wiman offered his lease scheme to the Montreal Telegraph. Allan refused to consider such a ludicrous proposal with a small, insignificant company from the west. Secretly, Allan hoped the offer he had submitted for the Dominion government to takeover and subsidize the MTC would occur, however, there was little hope due to the Pacific Scandal crisis of November 1873 in which Allan had played a pivotal role. On the other hand, the Dominion Telegraph through Western Union ownership had no choice but to accept the proposal because Allan could no longer sustain the MTCo's mounting debts.

Wiman suggested an alternative plan after the first plan failed. His new prospectus included leasing the lines of the two companies for which an annual payment would be paid by the GNWTCo. In return, he agreed to operate the entire two telegraph systems under his company, and move the head office from Winnipeg to Toronto. The proposal was taken to the Montreal and Dominion Telegraph directors and shareholders.

The Montreal Telegraph directors were outraged that their company could be obtained for so little. More to the point, it was a matter of principle not to be succumbed in such a humiliating fashion: "A company with a nominal capital [the GNWTCo], recently formed, with no one of standing in its make up to suggest the idea was too preposterous."[42] Without question Wiman stated, "... that unless his terms were accepted his Company, the Dominion and Western Union, behind them, would go into competition with them for Canadian business."[43]

At a combined meeting, shareholders of the MTC, GNWTCo and DTC "... approved amalgamation [in principal] with the Great North Western Telegraph Company [on or about 16 July 1881]."[44] Agreement was later reached with a yearly dividend to be paid by the GNWTCo to its new shareholders. The shareholder's of the Dominion Telegraph would receive "... six per cent. per annum, and to the Montreal Co.'s shareholders eight per cent. per annum."[45]

One month later, on 17 August 1881, the Montreal Telegraph Company, the GNWTCo., the Western Union and shareholders, under Great North Western control, "... undertook to operate and manage the telegraph lines and property of the Montreal Telegraph Company in Canada and the United States for a term of 97 years [sic - actually 99 years] from July 1, 1881...."[46] The recorded value of the Montreal Company was $2,207,533.25.

Immediately after acceptance, the GNWTCo restored the old tariff of 25 cents for ten words. The Western Union was no longer a competitor against the Montreal Company but now an ally with Western Union and the Dominion Telegraph Company. Indeed, a textbook case where the small company, in the right place at the right time, got the better business deal.

The Great North Western, with Western Union financial support, became Canada's largest telegraphic monopoly. The once great Montreal Telegraph Company was absorbed, however, it would retain its corporate name, property and financial account for many years.

The Great North Western introduced improvements to the telegraphic news service. As early as the 1850s, the telegraph was used to send news from the Atlantic coast to Montreal, primarily from Father Point, Quebec along British North American Telegraph Company wires. One circuit was for railway signalling and messages, commercial messages, and newspapers. Newspapers at this time depended strictly on hearsay and the undependable mail. Later, the newspaper's very existence depended on telegraph despatches. The merger of the Montreal and Dominion Telegraph Companies into the Great North Western, allowed for a special news circuit around 1882. This news circuit was established with the Associated Press and the GNWTCo's Press and Commercial News Department. Connections were made with newspapers established in Britain, Europe, United States and Canada.

The Herald.

VOL I. CALGARY, ALBERTA, JULY 2, 1885. No. 1.

TELEGRAPHIC.

FORT PITT.

STRAUBENZIE, June 29.— It was intended to make the start for home immediately on the arrival of Gen. Strange's column, but the programme has been altered, owing to two companies of the 65th (Montreal) battalion and one company of the Winnipeg Light Infantry having been left at various places between Calgary and Edmonton. These companies are now coming down the river on scows, and on the steamer Baroness, which left this afternoon to meet them. No move will be made until the steamer returns, which will be four days from this. The start will then certainly be made.

In a shooting match to-day the Royal Grenadiers, of Toronto, beat the 9th and Midland battalions.

MEDICINE HAT.

MEDICINE HAT, July 1.—The detachments here of the 66th, Halifax battalion, received orders this morning to report at Moosejaw en route to Winnipeg. The men were overjoyed at getting the order.

WINNIPEG.

WINNIPEG, July 1.—General Manager Van Horne gave the Winnipeg civic council no satisfaction in reply to its complaint against the new change of time table in the running of trains to the boundary.

At the public meeting called yesterday, to consider the means of raising money to erect a memorial to the volunteers, but two persons attended.

TORONTO.

TORONTO, July 1.—The investigation into the charges against Warden Massie, of the Central Prison, will be held on Thursday. Nicholas Murphy will represent the Catholic convention in reference to the charges preferred against Massie of ill-treating the Catholic prisoners.

HALIFAX.

HALIFAX, July 1.—The Scott Act has been carried in Guinsboro County. Only twenty Anti-Scott Act votes were polled.

NEW YORK.

NEW YORK, June 30.—Mrs. Dudley's trial, on the charge of attempting to shoot O'Donovan Rossa, has commenced. The New York Irish are exhibiting great interest in the result.

Fish, President of the Marine Bank, has been sentenced to ten years imprisonment.

LONDON.

LONDON, June 30.—Gladstone has addressed a letter to his Midlothian constituents, in which he intimates that he will again lead the Liberal party.

PARIS.

PARIS, June 30.—Rochford published this morning an Egyptian account of the death of Pain, in which the story is shown to rest on a runner whom the Mahdi subsequently killed.

Lord Salisbury is negotiating with the Porte for the occupation of Egypt by Egyptian troops to be paid by the Egyptian treasurer. This would involve an increase of twelve million pounds in the Egyptian loan. It is reported that Lord Salisbury's scheme includes the administration by England of the civil government of Egypt.

The Herald 2 July 1885, Calgary, Alberta - A typical front page example of how newspapers once relied heavily on telegraphic news from Canada and around the world. (R.G. Burnet)

It is interesting to note, "Every agent of the [GNWTCo] company was expected to report all local incidents of unusual importance, such as murders, robberies, deaths of prominent persons, fires, etc., and these being incorporated with the despatches received from the larger places, what was known to the news paper and telegraph fraternity generally as the 'Easson Service' was for many years a valuable help to many papers and the main reliance of not a few."[47]

The "Easson Service" refers to Robert F. Easson. He was born in Forfarshire, Scotland in 1838. At age eleven he began working for the Montreal Telegraph. In 1858, Easson was the sole agent at Father Point, Quebec for the Montreal Telegraph (originally the British North American Telegraph Company). This line was responsible for reporting shipping in the St. Lawrence but never succeeded. However, it did receive news despatches from Europe that Easson forwarded to the New York Associated Press. Around 1882, after the merger, Easson became Superintendent of Reports for the Press and Commercial News Department of the GNWTCo. In this position Easson streamlined telegraphic news releases, which became very profitable for the GNWTCo. In 1910, Easson retired after 52 years in the telegraph industry, and died shortly after. R.H. Hathaway called him the "Father of News Gathering in Canada." (By 1912, the Canadian Press formed and they broke their dependency on the GNWTCo, establishing their own independent circuit.)

Significantly, the GNWTCo did not receive any challenge to their monopoly until 1902. "The Great North Western Telegraph Company had constructed lines [by legal Agreement] along the route of the Canadian Northern Railway and operated those lines as a separate commercial telegraph business. The operating results between 1902 and 1915 were included as miscellaneous operation in the accounts of the Canadian Northern Railway Company."[48] The Canadian Northern Railway established their own telegraph company 15 May 1902, under 2 Edward VII, Chapter 51. Most charters now stated: "... [to] incorporated with powers to construct or acquire and operate overhead, underground, or cable lines of telegraph from and to any place in Canada or elsewhere, or to make connections with, or lease any telegraph lines in Canada or elsewhere...."[49] Because the GNWTCo had their lines along the Canadian Northern, in a quiet shareholder's takeover on 1 January 1915, the Canadian Northern Telegraph Company acquired the GNWTCo, by agreement from the Western Union. This acquisition gave Canadian Northern dominion over most national telegraph circuits except the Canadian Pacific Telegraph Company.

FIGURE FOUR

Principle Telegraph Route of the
Canadian Northern Railway - 1902
(Non-comprehensive)

Toronto	-Southern Route-	-Northern Route-
Washago	Longlac Junction	Longlac Junction
Parry Sound	Nipigon	Armstrong
Sudbury	Port Arthur	Superior
Capreol	Fort William	Sioux Lookout
Ruel	Atikokan	Kenora
Stackpool	Fort Frances	White
Oba	Rainy River	Ophir
Hornpayne	Winnipeg	Winnipeg
Longlac Junction		
	-from Winnipeg-	
	Portage la Praire	
	Brandon	
	Regina	
	Edmonton	
	New Westminister	

For ten years following this agreement, the GNWTCo and the Canadian Northern Telegraph had joint operational use of the telegraph lines. Railway signalling and business, commercial business and newspaper despatches were handled by the CNoR. Several times Canadian Northern tried to obtain a complete merger of railway, commercial and news service wires. However, "... as the two systems became more completely merged, losing their separate identities, the accounting as between the two became more difficult."[50]

At this time the Dominion Government was becoming more involved with railway and telegraph systems. The amalgamation of Canada's railways, except the Canadian Pacific and its holdings, was being formulated into the Canadian National Railways. Telegraphically, the same was also occurring. In 1918, the Canadian Northern, Great North Western Telegraph and the National Transcontinental Railway, were taken over by the government. The government then handed all telegraphic operations to the Grand Trunk Pacific Railway. Again, separate corporate names and accounts had to legally remain until the amalgamation process was completed.

FIGURE FIVE

Principle Telegraph Circuit of the Great North Western Telegraph, the National Transcontinental, the Grand Trunk Pacific and Canadian Northern Railway - 1918
(Non-comprehensive)

-New Brunswick-
Moncton
Pangburn
Chipman
McGivney
Napadogan
Summitt
Plaster Rock
Grand Falls
Edmunston
Pelletir
Picard

-Quebec-
La Pointe
Le Febvre
Monk
Bridge
Quebec
Port Neuf
Doheny
Fitzpatrick
Darey
Hibbard
Parent
Doucet

-Quebec-
Hughes
Cochrane
Driftwood

-Ontario-
Pullen
Mattice
Hearst
Grant
Kapita
Sioux Lookout
Armstrong
Graham
White

-Manitoba-
Ophir
Hazel
Winnipeg
Portage la Prairie
Rivers
Myra
Stenberg
Uno
Rocanville

-Saskatchewan-
Latimer
Cana
Raymore
Melville
Young
Saskatoon
Biggar
Oban

-Alberta-
Wainwright
Holden
Edmonton
Jasper

-British Columbia-
McBride
Prince George
Smithers
Hazelton
Prince Rupert

A drawing of the Canadian Northern Building. The bottom floor was for public use, such as sending/receiving telegrams, obtaining information and purchasing railway tickets. The top floor is believed to be the telegram/wire service floor. Floors two and three were likely CNoR business offices.

(CNR Public Affairs, #27471)

An example of the Government Telegraph station at Caribou. Most Government telegraph circuits were private holdings and did not provide railway communications. Like the Government Railways, a bankrupt or failing telegraph company was often taken over by the Dominion Government Telegraph until such time as it could be passed to another company or railway.
(National Archives of Canada, C67145)

By 1925, the Canadian Northern Telegraph Company owned 98 per cent of the Great North Western's capital stock. Typically, between 1916 and 1922, all income from telegraphic operations - primarily commercial and news services - was classified as 'Miscellaneous Accounts.'

IV.
CANADIAN NATIONAL RAILWAYS TELEGRAPH COMPANY

On 1 January 1921, an agreement was achieved between the Canadian National Telegraph Company and Canadian Northern Railway Company. Canadian National Telegraphs became the adopted trade name for telegraphic operational purposes. One year later, effective 1 January 1922 all facilities of the Canadian National Telegraph, the Grand Trunk Pacific Telegraph, and the Great North Western Telegraph Company of Canada were operated under one system and the trade name 'Canadian National Telegraph.'[51] However, the CNT trademark existed one year earlier, because, on 17 September 1920, the Canadian Northern Railway Telegraph Company name became the "Canadian National Telegraph Company" (Dominion Order in Council P.C. 2247). Two weeks later on 1 November 1920, management of the Grand Trunk Pacific Telegraph Company ended, being transferred to Canadian National Telegraphs (CNT). It would take another 35 years before all the legal wrangling of Company names and their rights, properties, previous contracts and stocks passed completely into Canadian National Railways' control.

Canadian National Telegraphs moved to formulate their company objectives in 1921. Essentially what had been built by the amalgamated companies would only be maintained and slightly expanded by CN Telegraphs now. Although there was construction of new lines it was minimal. Finances would remain under the original company names with separate accounts from CNT as established by the agreements. This gave the companies specific legal rights under their lease. For CNT to assert Canadian wide operating control, they had to obtain Western Union's properties. Therefore, on 1 September 1924, Canadian National Telegraphs began purchasing specific property in British Columbia from Western Union. The Western Union services in Vancouver, Victoria and New Westminister along with their property, equipment and facilities, became part of CNT for $97,618.66.

Under Canadian National Telegraphs, the Dominion Telegraph Company was reestablished as a holding company. This was CNT's first major move towards complete financial and operational control across Canada. Dominion Telegraph Securities Company Limited was created on 12 January 1925. Property, equipment, poles and wires, income from commercial telegraphs, news service income and all other assets dating from 12 June 1879, were placed under this company. Dominion Securities had no assets of their own. Their only right was to claim the annual rental of $62,500.

Three days later, on 15 January 1925, an Agreement was signed between the Dominion Telegraph Company, American Union Telegraph Company, Western Union, Great North Western Telegraph, and the Dominion Securities Company "... in consideration of the payment to them by the Great North Western of $116,640, released the American Union, the Western Union and the Great North Western from their respective covenants [as successive lessees] to return to Dominion Telegraph Company at the expiration of the original 99 years lease term the telegraph lines and facilities of the Dominion Telegraph Company in working order and repair ... [and] ... upon termination of the 99 year term, to transfer to the Great North Western all of their right, title and interest ... west of the Province of New Brunswick and to transfer to Western Union all said rights in respect of the balance of the lines ...."[52] Separate accounts would be effective until 31 December 1927. On 1 January 1928, "... the accounts were consolidated under the name 'Canadian National Railways.'"[53] Canadian National Telegraphs was as aggressive as the Montreal Telegraph 40 years earlier.

On 26 April 1927, Canadian National inaugurated the Carrier Current System. Briefly, Canadian National and Canadian Pacific Railways were entering a period of technological firsts. The competition between the CNR and CPR to have headlining technological 'first' was very important. Canadian National developed carrier current from the telegraph-telephone combination. It began as a ten-channel system and could theoretically be expanded to 160 channels. The telegraph and telephone could share the same wire without interference or loss. (This is further developed in Chapter Four.)

On 1 July 1929, W.D. Robb, Vice-President in charge of Canadian National's Telegraph and Telephone Department, and W.G. Barber, General Manager, made a significant announcement. On this date, all Western Union services, facilities, properties and connections with Europe, United States and the Maritimes, became part of CNT. This gave CNT an additional 6670 miles of wire and 188 offices in New Brunswick, Nova Scotia and Prince Edward Island; Newfoundland joined in 1949, but currently was under the Western Union, Anglo-American Telegraph and Newfoundland's Department of Posts and Telegraphs. The Western Union agreement included the outright purchase of all assets, while the Western Union reserved "... ownership of all pole lines, wires and appurtenances and the cable and cable stations and equipment and other property necessary for its cable business."[54] This agreement gave CNT a staggering telegraph system to go with their railway. They now operated 159,736 miles of wire and 2131 telegraph offices. In short, "Telegraph history is written in the demand the public makes for service. It is a demand that brooks no obstacles...."[55] And the people of Canada, and businesses, were demanding it.

The Depression Years slowed Canadian National's opportunity for acquisition and expansion. Not only did the railway suffer, telegraph business and commercial messages declined. It is noteworthy, that the government created a serious defect in the formation of the Canadian National structure generally. Canadian National was not required to raise new capital. The primary goal established under Henry Thornton was for the CNR to remain independent of government. However, when the CNR borrowed money, a government guarantee with the money was understood. Canadian National neither had to maintain its earnings on their capital nor seek new market capital or investments. This was especially noticeable during the Depression. In later years this imperfection would become more dramatic.

It was eighteen years later, after the Depression and World War Two, before Canadian National Telegraphs started profiting again. November 1, 1948, was a celebrative day for the CNR and CNT. On this date they acquired two long time telegraph companies. The first Agreement was the purchase of all legal rights and property of the Dominion Securities Company. Their lease was obtained for $1,017,900.41 after being approved on 23 May 1949 by Dominion Order in Council P.C. 2600. However, it was not until 5 December 1949 that most interests were transferred. A further property and payment Agreement was reached on 14 February 1950. Finally, Dominion Securities "... surrendered its charter and the corporate existence of the Company was terminated as of 3 April 1950...."[56]

The second company acquired on 1 November 1948, was the Great North Western Telegraph Company. It transferred "... all its rights, title and interest in the Agreement of January 15, 1925...."[57] to the Canadian National Railway Company.

Less than a year later, on 1 April 1949, the CNR obtained Newfoundland's Department of Posts and Telegraph upon their entry into Canadian Confederation. The CNR would need to renovate the Island's telegraph systems. Then, on 1 March 1950, Canadian National purchased all facilities, property and the Anglo-American Telegraph Company's lease held by Western Union since 12 March 1912. The purchase amount was $39,371.

It is interesting to note, the Montreal Telegraph Company was not completely absorbed until 1954. This company had not operated officially since 1882 because of the terms of its lease to the Great North Western, it remained a corporate entity with some property still under its name. In 1944, the MTCompany wanted to sell its remaining building and land in Toronto for $20,000. The Great North Western Telegraph Company as Guarantor agreed with the sale. However the MTCompany wished to keep the money and divide it up among the remaining shareholders. The GNWTCo would not allow it, so they took the MTCo to court.

On 1 February 1954, Canadian National offered the Montreal Telegraph $3 million. This was for the purchase of all rights and interests for its original 99 year lease it signed with the GNWTCo. In the courts, CN won on the condition the "... Montreal Telegraph Company be dissolved as provided for in Cap. 93, Statutes of Canada 1882."[58] On 20 April 1954, agreement was reached with the remaining shareholders. Then, on 25 May 1954, the Montreal Telegraph transferred all remaining properties from the original lease of 17 August 1881 and 1 June 1883 with the GNWTCo to the CNR. After a rental payment of $27,500 from 1 April to 31 May 1954 was made, the Montreal Telegraph Company ceased to exist as of 1 June 1954.

In a more amicable arrangement, the final remnant of the Grand Trunk Pacific Telegraph Company was arranged on 13 May 1954 by Dominion Order in Council P.C. 1954-714. After a counterpart

Agreement was filed on 8 June 1954, all properties, shares, stocks and interests were turned over to the CNR. The Grand Trunk Pacific Telegraph Company ceased as a corporate entity now.

Thirty-five years after the descriptive title, "Canadian National Telegraphs" was legally adopted, the "Canadian National Telegraph Company became a wholly owned subsidiary of [the] Canadian National Railway Company ... [on] June 11, 1956."[59]

V.
THE TELEGRAPH IN THE MARITIMES 1846 TO 1950

It is significant to highlight telegraphic progress in the Maritimes during the same period as the Montreal and Dominion Telegraph Companies development. As Upper and Lower Canada began their telegraphic system in 1846, the Maritimes were developing a very different type of communication network. Although economic trade with the United States from the Maritimes was condemned, it did play "... a large part in the commercial affairs of the Maritime Provinces ... and the people of the Maritimes continued to look for trade to the United States ..."[60] and the West Indies and England. This Maritime-American economical trade route was much preferred to one with Upper and Lower Canada.

Originally, telegraphic construction was "... in response to the demand for telegraphic transmission to New York of European news received by steamers arriving at Halifax."[61] The Press Association of the United States established a two-part communication system. This service began in 1847 between Halifax and Digby, Nova Scotia to Portland, Maine. It entailed the transport of bulk goods, trade articles and passengers, but the highest priority was news despatches from Europe to Halifax by steam ship. If the despatches from Europe went directly to New York by ship, it would be several more days before the information was received and published in the newspapers. Halifax, therefore, provided the best and fastest access for transferring information because of its proximity to Portland, Maine.

The first Nova Scotian Act to construct the telegraph line between Digby, Nova Scotia and Portland, Maine was revoked about 1847. Later that year, a second Act was drawn up and passed "... authorizing the appointment of five commissioners to erect a line to the border until it met with the N.B. [New Brunswick] line to be known as 'The Acadian Telegraph'."[62] F.N. Gisbourne (B: 8 March 1824 / D: 30 August 1892), who owned a Quebec-based telegraph construction company and would play a major role in the Maritimes telegraphic development, was contracted to build the circuit through Nova Scotia and New Brunswick to Portland, Maine.

Gisbourne's company began construction on 4 October 1848 from Calais, Maine to Amherst, Nova Scotia. The New Brunswick Electric Telegraph Company was incorporated at this time. Capital Stock was established at $40,000; shares were offered at $40 each. The first leg of the circuit was completed in about two months.

From Amherst to Halifax the second phase of the circuit would take slightly longer to build. Construction began on or about 1 January 1849. By 9 November 1849 the circuit was completed linking Halifax through to Portland, Maine and the American Telegraph Company. The official opening of the Halifax-Amherst-Portland circuit occurred on 15 November 1849, with the arrival of the first telegram in Halifax from St. John. It was reprinted in the Halifax newspaper, and simply read: "Great Storm at Portland - Saint John, November 15 - 10 buildings were blown down on Friday night. No other news per steamer."[63]

It is interesting to note that complementing the ship service before the telegraph circuit was the Canadian version of a Pony Express. The first recorded express horseback ride was 21 February

1849, between Halifax and Digby. Under ideal conditions, the rider covered the 150-mile distance in eight-and-half to nine hours. The express bag was transferred to a small boat that carried the news to a ship in the harbour. The ship then set sail for Portland. On arrival at Portland, the Press releases were telegraphed to New York. However, it was the Associated Press' eagerness to obtain news despatches faster that influenced telegraphic communication in the Maritimes. Although the use of a horse rider and connecting ships "... was expensive ... the Associated Express agents, offered to guarantee the payment of a liberal subsidy to any Company who would construct a telegraph between Calais, Maine and Halifax, which could be available for press matters."[64] Seeing an economic opportunity, the Nova Scotia Government agreed to finance the construction of a telegraph circuit to replace the express rider. Horse rider service ended November 1849.

FIGURE SIX

Principle Telegraph Route of the Calais to Halifax Circuit - ca1850 (Non-comprehensive)

Calais, Maine	Moncton	Shubenacadie
St. Stephen, NB	Memramcook	Milford
St. George	Dorchester	Elmsdale
Bennfield	Sackville	Enfield
Lepreau	Amherst, NS	Wellington
St. John	Maclan	Beaver Bank
Martin	Springhill	Waverley
Hampton	Londonderry	Bedford
Norton	Belmont	Halifax
Sussex	Hyde	
Salisbury	Brookfield	
Petitcodiac	Alton	

The Nova Scotia government operated the Halifax to Amherst circuit. But the "... opposition party in the [Nova Scotia] legislature, ... felt it was wrong and dangerous to have the telegraph under governmental control."[65] As a result, in 1851, a private company was established and called the Nova Scotia Electric Telegraph Company. For several years it would operate under a government cloak. By 1855, the Nova Scotia Electric and the circuits in New Brunswick, became a financial failure; this would give the American Telegraph Company its first opportunity to take over Maritime communication.

Although Gisbourne had won two telegraphic contracts in New Brunswick and Nova Scotia, he was not successful in securing a direct and united telegraph route from Quebec through the Maritimes. The British North American Telegraph Company, incorporated in 1847 to connect the Maritimes with Quebec and advise of shipping movements and news despatches, was stopped at Riviere-du-Loup by the New Brunswick Legislature. The completion of this circuit was viewed as a threat to the preferred trade and more profitable contacts with the United States. Whereas the telegraph helped to cement relations between Quebec City, Montreal, Toronto and Buffalo, "... there was not a 'considerable' increase in commercial activity due to the telegraph ..."[66] between Riviere-du-Loup, St. John, Fredericton and Halifax. However, within their own boundaries and through the interests of the private sector, funds were raised and a circuit completed on 30 April 1851 between St. John and St. Andrew's New Brunswick. (Not for another twenty-five years, when the Intercolonial Railway was opened, did Ontario, Quebec and the Maritimes have a through Canadian circuit.)

In 1852, Gisbourne constructed the first North American submarine telegraph cable. This cable, completed Monday 22 November 1852, linked Cape Tormentine, New Brunswick to Carlton Head, Prince Edward Island. Interestingly, with the success of the Maritime submarine cable, Gisbourne approached Cryrus W. Field, president of the American Telegraph Company in 1854 with a Trans-Atlantic submarine cable plan. Gisbourne's plan was to build this cable under the Atlantic Ocean. Field liked the potential economic possibilities so much he purchased Gisbourne's trans-Atlantic plans. This move, coupled with the financial failure of the Nova Scotia Electric and New Brunswick Electric, launched the long term American control and ownership of the Maritime telegraph network.

In 1856, an agreement with the American Telegraph Company was reached. This agreement included most telegraphic lines in New Brunswick and Nova Scotia. Further, a lease of ten years with renewal rights and an annual rental fee of $3,000 would be paid to the provincial government. By 1860 the American Telegraph leased the Nova Scotia Electric Telegraph circuit. It was agreed an annual rental fee of $6,500 would be paid plus a renewable clause for ten years.

Maritime telegraph circuits were obtained by the American Telegraph Company in anticipation of the trans-Atlantic cable. The American Telegraph Company wanted to have complete control of the Maritime lines.[67] Newfoundland would provide an excellent terminal station for the future trans-Atlantic cable. As early as 6 May 1854, Field established the New York, Newfoundland, and London Telegraph Company to provide another ATC circuit from Newfoundland through Halifax to New York. Until the New York, Newfoundland and London Telegraph Company was fully operational, a boat connection with the mainland was used. Along the proposed submarine cable route, a twice monthly whale boat intercepted the mail boat. The boat crew picked up sealed containers of mail and telegraphic dispatches of commercial and news medium. By 10 November 1856, this telegraph-cable system was operating from St. John's Newfoundland to Halifax Nova Scotia to New York, ending the boat services.

Established by the New York, Newfoundland, and London Telegraph Company was the first telegraph station at Bulls Arms on the shore of Trinity Bay. The circuit then followed the south shore of the island to Come-By-Chance, Pipershoal, La Manche, Long Harbor, Bay d'Espoire, Bay du Nord, White Bear Bay, Granny's Brook, Garia, Rose Blanche, Channel, and Cape Ray. From Cape Race, a cable was submerged across the Gulf of St. Lawrence to Cape North on Cape Breton Island. This telegraphic cable connected the land line telegraph in Halifax, then through Digby, into Portland, Maine then Boston and New York.

The first trans-Atlantic submarine cable construction began between 7 and 11 August 1857 from Ireland to Trinity Bay, Newfoundland. But when tested, the cable completely failed. A second submarine cable was attempted in 1858 but failed in October 1858 due to insulation eroding off the cable. On 27 July 1866, the third submarine cable from Great Britain operated successfully. The ships Eastern, Medway and Albany arrived off Heart's Content Harbor, joining the two continents by cable. In 1866 as well, the American Telegraph Company leased to Western Union, control of their telegraphic lines in the Maritimes that included the "Acadian Telegraph", the Nova Scotia Telegraph and New Brunswick Telegraph companies.

Interestingly, the precedent set for the telegraph failed for the railway. Joseph Howe, the Nova Scotian premier, made every effort to keep trade moving along the Atlantic coast. By September 1850, Howe offered a resolution that would make the construction and control of highways in Nova Scotia a purely governmental objective. His assumption was based on the 1849 Guarantee Act that provided public aid for the second half of any road that was at least seventy-five miles or more in length. Howe knew that millions of English pounds sterling would be required to initiate railways and this in all likelihood could be obtained from the Crown. However, such construction was left to the private sector and only small grants or guarantees might be forthcoming. Howe asked the mayor of Halifax to deliver his petition for road, highway and railway development to Sir John Harvey, the

Lieutenant-Governor. Harvey then conveyed the same to the colonial secretary, Earl Grey, based on a previous precedent allowing for the "... construction of the electric telegraph, which not only connects Halifax with the chief towns of New Brunswick and the State of Maine, but forms the most important link in the chain of communication between the old world and the new, and the success which has attended the appropriation of the public funds, has attracted public attention to the practicability used importance of placing the railroad beside the telegraph."[68] The Crown answered negatively in October 1850, as telegraphic communication was far more significant than seasonal railway operations.

Joseph Howe's next railway construction attempt occurred in 1853 when he initiated the Truro to Pictou railway line. It would take fourteen years to complete this 52-mile long railway because of financial and political jostling. By October 1865, the railway was still unfinished so Premier Tupper of Nova Scotia placed Sandford Fleming in charge of construction. Fleming had not built a railway until now. He enthusiastically began the challenge, improving on roadbed, bridges, tracks. Fleming included the telegraph along the entire length of the route and always well ahead of track laying crews. This allowed him to be constantly abreast of daily end-of-track locations, supplies and train movements to complete the work. The single circuit design, like the Nova Scotia Telegraph and New Brunswick Telegraph, was one wire transmission along galvanized iron wire. The railway line was completed by the middle of October 1867.

FIGURE SEVEN

Principle Telegraph Route of the Truro and Pictou Circuit - ca1866 (Non-comprehensive)

Truro-Valley-Hopewell-Stellarton-New Glasgow-Westville-Pictou

With attitudes changing towards the United States, many hailed the coming of the Intercolonial Railway. George Brown, speaking about the railway and communication, stated on 8 February 1865: "I am in favour of this union because it will give us a sea-board at all seasons of the year. It is not to be denied that the position of Canada, shut off as she is from the sea-board during the winter months, is far from satisfactory ... [and] ... The Maritime Provinces are equally cut off from communication inland."[69] John A. Macdonald appealed for a united Canada that included the Maritimes. A new Dominion would also embrace New Brunswick and Nova Scotia as one economic unit. A local self-interest economy would not work. The Intercolonial Railway would be the way to bond the Maritimes with the rest of Canada. The railway would make another strong stand against American Imperialism. Hinting at telegraphic communication, Macdonald stated:

> In the case of a union, this railway must be a national work, and Canada will cheerfully contribute to the utmost extent in order to make that important link, without which no political connection can be made. What will be the consequence to this city [Halifax], prosperous as it is, from that communication? Montreal is at the moment competing with New York for the trade of the great West. Build the [rail] road and Halifax will soon become one of the great emporiums of the world.[70]

The British North America Act of 1867 Section 145 specifically dealt with the Intercolonial Railway.

This clause was politically a major concern for the Maritime inclusion: "... the Intercolonial Railway is essential to the Consolidation of the Union of British North America and to the Assent ... of Nova Scotia and New Brunswick...."[71] (Section 145 was repealed in 1893 as it was no longer necessary in the BNA Act.) Section 92 of the BNA Act provided exclusive legislative powers to all provinces, to make laws in matters concerning them directly, however, Subsection 10 under Local Works and Undertakings excluded "... railways [and] telegraphs ... extending beyond the limits of the province...."[72] In essence, should a nationwide crisis occur, the Prime Minister required nationwide communications. Macdonald had further legislation where, in such an event, he could use the telegraph circuits to alert the nation. Provincial circuits were simply not secure without a Federal mandate.

The Intercolonial Railway was completed in 1876 from Riviere-du-Loup to Halifax. Sandford Fleming played a major role in its operation and laying the foundations for telegraphic unity as well. There now existed direct telegraphic communication between Toronto, Ottawa, Montreal, Quebec, Riviere-du-Loup, St. John, Halifax, Britain and the United States.

FIGURE EIGHT

Principle Telegraph Circuit Route of the Intercolonial Railway (Maritimes) - ca1876 (Non-comprehensive)

-Nova Scotia-
Halifax
Windsor
Annapolis
Truro - branch to
Pictou
Amherst
Painsec

-New Brunswick-
Moncton - branch to Salisbury,
Petitcodiac, Norton,
Hampton, St. John
Shediac
Chatham
Bathurst
Campbellton

-Quebec-
Matapedia
Clark Brook
Routhierville
Sayabec
Mont Joli
Father Point
St. Simon
Trois Pistoles
Isle Verte
Riviere-du-Loup
Saint-Andre
Saint Charles
Quebec
Drummondville
Montreal

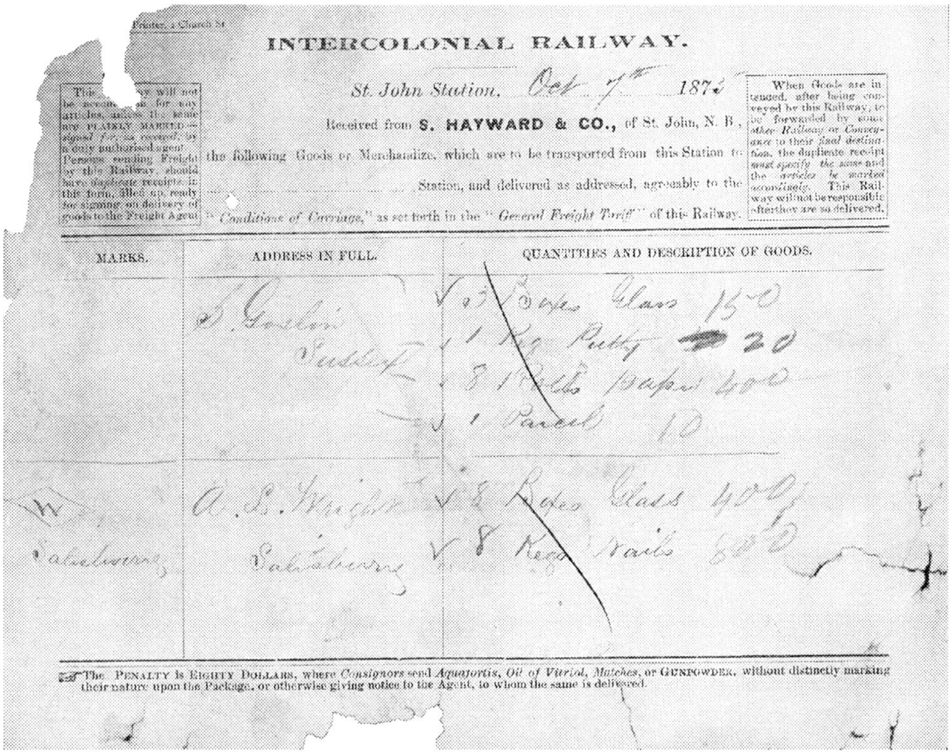

INTERCOLONIAL RAILWAY.

St. John Station, Oct 7th 1875

Received from **S. HAYWARD & CO.**, of St. John, N. B., the following Goods or Merchandize, which are to be transported from this Station to .... Station, and delivered as addressed, agreeably to the "*Conditions of Carriage*," as set forth in the "*General Freight Tariff*" of this Railway.

This Railway will not be accountable for any articles, unless the same are PLAINLY MARKED — *signed for*, as *received*, by a duly authorised agent. Persons sending Freight by this Railway, should have duplicate receipts in this form, filled up, ready for signing on delivery of goods to the Freight Agent.

When Goods are intended, after being conveyed by this Railway, to be forwarded by some *other Railway or Conveyance* to their *final destination*, the duplicate receipt *must specify the same* and the *articles be marked accordingly*. This Railway will not be responsible after they are so delivered.

MARKS.	ADDRESS IN FULL.	QUANTITIES AND DESCRIPTION OF GOODS.
	S. Goslin Sussex	5 Boxes Glass 150 1 Keg Putty 20 8 Rolls paper 400 1 Parcel 10
W Salisbury	A. L. Wright Salisbury	2 Boxes Glass 40 8 Kegs Nails 800

The PENALTY is EIGHTY DOLLARS, where *Consignors* send *Aquafortis, Oil of Vitriol, Matches*, or GUNPOWDER, without distinctly marking their nature upon the Package, or otherwise giving notice to the Agent, to whom the same is delivered.

Two examples of waybills from the Intercolonial Railway, respectively 7 October 1875 and 24 April 1889. Of interest are the varied contents, prices and improved graphics on the second bill. If there were problems with contents, damage, late shipment or costs, the query would be telegraphed to the source to be remedied. (R.G. Burnet)

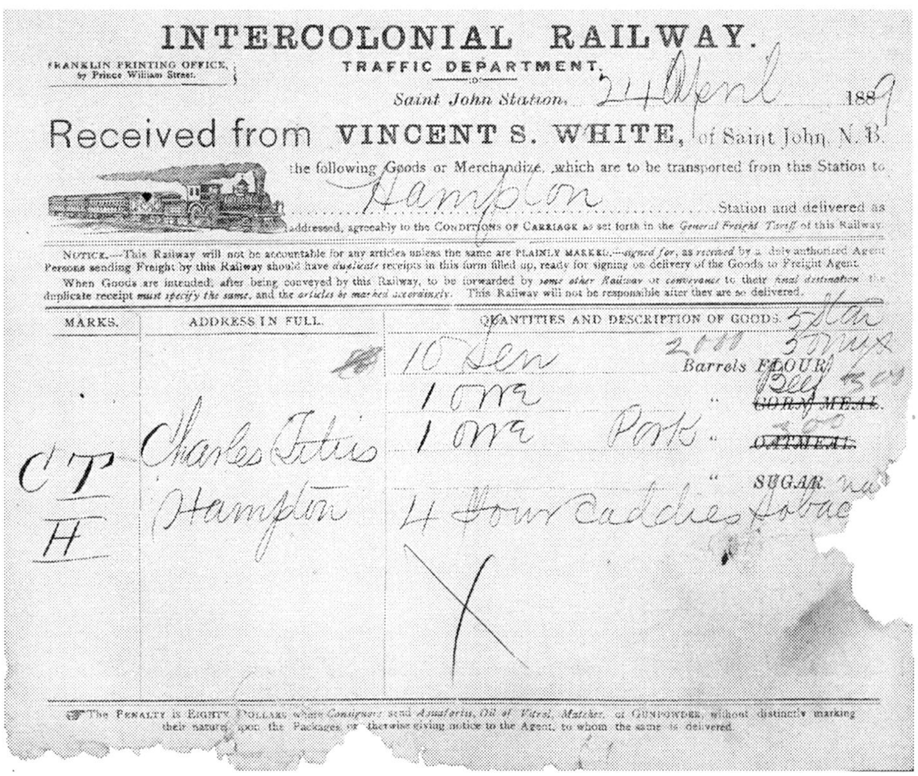

INTERCOLONIAL RAILWAY.

TRAFFIC DEPARTMENT.

FRANKLIN PRINTING OFFICE, by Prince William Street.

Saint John Station, 24 April 1889

Received from **VINCENT S. WHITE**, of Saint John, N.B.

the following Goods or Merchandize, which are to be transported from this Station to Hampton Station and delivered as addressed, agreeably to the CONDITIONS OF CARRIAGE as set forth in the *General Freight Tariff* of this Railway.

NOTICE.—This Railway will not be accountable for any articles unless the same are PLAINLY MARKED,—*signed for*, as *received* by a duly authorized Agent. Persons sending Freight by this Railway should have *duplicate* receipts in this form filled up, ready for signing on delivery of the Goods to Freight Agent.

When Goods are intended, after being conveyed by this Railway, to be forwarded by *some other Railway or conveyance* to their *final destination*, the duplicate receipt *must specify the same*, and the *articles be marked accordingly*. This Railway will not be responsible after they are so delivered.

MARKS.	ADDRESS IN FULL.	QUANTITIES AND DESCRIPTION OF GOODS.
C T / H	Charles Titus Hampton	10 Sen 2.00 5 Star Barrels ~~FLOUR~~ 3 mix 1 one Beef ~~CORN MEAL~~ 5.00 1 one Pork ~~OATMEAL~~ SUGAR 4 Four caddies tobac

The PENALTY is EIGHTY DOLLARS where *Consignors* send *Aquafortis, Oil of Vitriol, Matches*, or GUNPOWDER, without distinctly marking their nature upon the Packages, or otherwise giving notice to the Agent, to whom the same is delivered.

The Western Union Telegraph Company of the United States purchased almost all telegraph lines in Nova Scotia for $150,000 in 1872. Western Union, which had been incorporated in 1851 at Rochester, New York as the New York Mississippi Valley Printing Telegraph Company and renamed in 1856 to the Western Union Telegraph Company at Ezra Cornell's suggestion, did not have complete control. The Montreal Telegraph Company owned the circuits from Campbellton, New Brunswick to Sackville denying complete WU communication dominance.

The 1874 Dominion Government finally realized the potential of the Maritimes for communication, trade and Canadian unity. The Electric Telegraph Act of 1852 specifically dealt with Ontario and Quebec only. In 1867 the Act was revised under the General Laws of the Confederation of Canada but still excluded the Maritimes. Becoming slowly aware of American influence in the Maritimes, the legislature revised the Act to collectively include the entire Dominion and give the Maritimes the same legal rights for telegraphic charters, but not until 1874. In short, whereas the telegraph did unite Upper and Lower Canada commercially, the same cannot be said about the Maritimes to the rest of Canada until much later. Politically, it was a belatedly vain attempt to exclude the already long established American telegraphic companies.

Another outside company in the Maritimes was the Anglo-American Telegraph Company. This British corporation established a telegraphic monopoly in Prince Edward Island (PEI) in 1852 and Newfoundland in 1854. Not until 2 October 1871 however, did PEI begin railway construction, following the established telegraphic route of the Anglo-American from Alberton through Summerside and Charlottetown on to Georgetown, Tignish, Souris, St. Peter's Bay along with many branch circuits. The PEI railway was completed 12 May 1875.

FIGURE NINE

Principle Telegraph Circuit of the Anglo-American Telegraph Company on Prince Edward Island - ca1852 (Non-comprehensive)

Tignish	Charlettetown	Montage
Alberton	Mount Stewart	Georgetown
Summerside	Vernon	Souris
Bordon	Marry Harbour	Elmira

In Newfoundland, the telegraph began in 1854 but not until 9 August 1881 did railway construction between St. John's and Hall Bay begin. This railway followed the established Anglo-American telegraph circuit, which, by now, was operating under a Western Union lease. The circuit went from St. John's to Hall's Bay, South West Arm, Random Sound, Clode Sound and across the Exploits River. The early contractors in Newfoundland however, did a haphazard job. In 1890, Robert Gillespie Reid of Montreal was contracted to build, rebuild and revitalize the substandard work previously completed. Railway and telegraph work was completed in 1897 with the Anglo-American operating under a renewed Western Union lease.

In 1898 the Newfoundland Tory Government accepted Reid's earlier economic development plans for Newfoundland. Part of the plan included Reid operating and controlling the railway until 1948, when it would become the property of his successors. The plan included the takeover of "... the island's thousand-mile telegraph system for $125,000 and operate the service for fifty years. Until 1904, when the Anglo-American Telegraph rights expired, [Reid] would receive a $10,000 annual subsidy."[73] However, in 1900 the Liberals won the election. Their first objective was to modify Reid's 1898 terms. For compensation of one million dollars, Reid agreed to surrender his telegraph control along with several other land and railway interests. This was acknowledged in a contract dated 2 August 1901. The telegraph was therefore returned to the Newfoundland government. The Liberals did amend the agreement so that Reid would operate the telegraph facilities and "... be compensated for upgrading the system."[74]

On 12 March 1912, The Anglo-American obtained a renewal to the Western Union lease. But there was considerable doubt to how long this company and Western Union would stay in the Maritimes and operate unprofitable circuits. By 1923, the Newfoundland government gained control of the telegraph network under the island's Department of Posts and Telegraphs. Western Union operated chiefly out of St. John's with leased facilities under Anglo-American.

FIGURE TEN

Principle Telegraph Circuit of the Anglo-American/Western Union Telegraph Company on Newfoundland - ca1890 (Non-comprehensive)

Port aux Basques	Bonnavista
Cape Ray	Shoal Harbour
Stephensville	Placentia
Corner Brook	Argentina
Notre Dame	St. John's
Terra Nova	

As Canadian National telegraphic strength grew nationwide, Western Union and Anglo-American realized it would be better for another company to operate an unprofitable system. And Canadian National Telegraphs was eager to obtain a Maritime monopoly. On 4 January 1929, an agreement reached between the CNR, Western Union and Anglo-American, stated that Western Union "... reserved ownership of all pole lines, wires and appurtenances and the cables and cable stations and equipment and other property necessary for its [WU] cable business...."[75] Canadian National purchased all other property such as land, rights, easements for $1,103,295, which included $126,788.00 for the purchase of all Anglo-American property left in PEI and New Brunswick.

Not until 1 April 1949, could the CNR provide its service to Newfoundland. Prior to CNR operational control, the Department of Posts and Telegraphs operated the system with Western Union's cable business operating out of St. John's. "Both parties believed it would be unprofitable and undesirable for both parties to continue providing telegraph service in Newfoundland and agreed that Canadian National Railways Company should take over operation of service, theretofore performed by Western Union."[76] By conveyance, the CNR, Anglo-American and Western Union agreed that only Canadian National Telegraphs would operate telegraphic service in Newfoundland. Payment of $45,000 was made to Western Union. The agreement was reached on 13 December 1950 but was made retroactive to 1 March 1950. The CNR now operated a successful telegraph system in every Canadian province.

CHAPTER TWO

THE FORMATION AND DEVELOPMENT OF THE CANADIAN PACIFIC TELEGRAPH COMPANY 1860 to 1937

I. THE TELEGRAPH FOR CANADA'S WEST

From the time Upper and Lower Canada had formed until the early 1860s, the 'chartered known west' was no further than to the top of Georgian Bay. The 'far west' was everything else owned by the Hudson Bay Company to the Pacific Ocean. There were no large settlements past Fort Gary and certainly nothing of consequence between the Fort and Toronto. Economically, there was no reason to push west at this time.

The United States helped to spur Canada into an independent nation. In the 1840s, the manifest destiny of the United States was to rule North America. Mexico though, bore the major assault. Shortly afterwards, the American Civil War erupted placing the Province of Canada in two difficult positions: the Fenian Societies, and persistent British-American conflicts.

In 1858 the first significant movement to include the Prairies came. The North-West Transportation, Navigation and Railway Company had been incorporated with "... powers to establish communication from one or more points on the shore of Lake Superior to any point in the interior within the limits of Canada."[1] One year later an amended Act was issued changing their original name to The North-West Company with additional power to construct a telegraph line. However, this Act also expired through non-use.[2] In 1862 the North-West Company was authorized a third time to "... establish communications within the Northern and Western limits of Canada."[3] Not surprisingly, little was accomplished.

A formerly planned railway and telegraph route west began in early 1860. Edward Watkin, (B: 26 September 1819 / D: 13 April 1901) a prior manager of the Manchester, Sheffield and Lincolnshire Railway, was hired by Glynn, Mills and Baring Brothers Company to solve the Grand Trunk Railway's financial troubles. In 1860, Watkin recommended that a railway route must go from the Atlantic to Pacific Oceans. He proposed the original idea of an Intercolonial Railway in the Maritimes, but "...we say then, 'Establish an unbroken line of road and railway from the Atlantic to the Pacific through British territory'. Such a great highway would give shorter distances by both sea and land, with an immense saving of time ...."[4] Watkin, an effective troubleshooter, reported that the GTR could earn higher profits if they built their tracks and a telegraph circuit to encompass the Maritimes, Upper and Lower Canadas, the Prairies and into British Columbia. However, the Grand Trunk Railway was more interested in protecting their railway interests from Sarnia to the east. After analysing the Grand Trunk company plans, Watkin concluded that the GTR was an "organized mess - the sink of iniquity."

Watkin and the Duke of Newcastle (the Colonial Secretary), began a series of meetings in 1862. Watkin proposed three fundamental objectives. The first was to organize a telegraph company. Next, establish the Intercolonial Railway for the Maritimes. Finally, a plan was needed to acquire Rupert's Land from the Hudson Bay Company (HBC). Watkin genuinely wanted to improve the financial profits of the GTR; Newcastle needed to gain secure political stability and create a single united, unified Canadian dominion. Their goals, attitudes and dreams were fuelled by troubles south of the border. An all-Canadian railway and telegraph system would conveniently establish a strong communication link and barrier from the Atlantic to the Pacific Oceans.

Watkin "... incorporated the Atlantic and Pacific Transit and Telegraph Company to bridge the gap of 2,500 miles between Canada and British Columbia with a means of communication, as a tangible token of unity. But he had found little local support for such an undertaking; it was beyond the thought range of Canadians [at this time]."[5] To him, this was in the best interest of the Grand Trunk Railway and Canadian confederation.

In the Spring of 1862, the government decided that a "...telegraph line, a cheaper, faster way of establishing sovereignty over western British North America, was the short-term answer."[6] The British Government would only support expenditures for imperial economic and political gains. At this time, a telegraph system across Canada was considered an unattainable concept. Watkin was intuitive enough to know the Canadian government had to take immediate action with the HBC because American soldiers and settlers were beginning to colonize British Columbia and the Prairies. Watkin took his plans to Glynn and Baring Brothers at the Colonial Office. His proposal of 11 December 1862 stated: "... that upon a Governmental guarantee of interest at the rate of 4 per cent. a sum of five hundred thousand pounds would be immediately raised for the purpose of constructing at once a telegraph line and a common highway for carrying the mails and the traffic between Canada and the Pacific."[7] Watkin initially suggested a joint railway and telegraph line to the Pacific through Canada. Being a young colony striving for an independent Dominion, he believed money could be obtained from Great Britain more easily. However, he discovered that he could not expect to obtain much, if any money.

The British bankers at Baring became concerned. Privately, Thomas Baring and George C. Glynn did not want an alliance with Watkin and Newcastle for an all Canadian railway. Baring believed Watkin was being manipulated by the government and would become culpable in the geographic and political chess match where no winner or compromise would emerge. The bankers felt trapped by the estimated construction plan costs to build west and to build the Intercolonial Railway in the Maritimes. Watkin was aware that the Canadian and English governments would not invest any finances to colonize the west. Watkin turned to the private sector of British financiers and industrialists who were eager to support a joint railway-telegraph system. Because of private financial investments, Watkin received consent to establish the Atlantic and Pacific Transit and Telegraph Company. This company was charged with bridging the three thousand or so miles from Halifax, Nova Scotia to New Westminster (British Columbia's colonial capital). A road followed by a railway line would come next. A telegraph and railway connection would provide a strong signal to the Americans that Canadian and British unity was serious.

Rupert's Land needed to be acquired from the Hudson Bay Company. Watkin's Transit and Telegraph Company requested a stretch of land ten miles wide across Rupert's Land before it could even consider commencing construction. The proprietary order of construction, however, was altered. The ten mile wide swath of land would provide enough land for a roadway first then a railway and telegraph right-of-way with room to expand facilities. The building of settlements would follow as required. He estimated that settlements would reach the Fraser River in about thirteen years. Next, the railway would replace the path but not until the population out west had grown to warrant its need. The Hudson Bay Company fought hard to abort any type of settlement on their Prairies as colonization meant destruction of their original land charter granted from England. However, the HBC was aware as early as 1851 their land might be parcelled off. Further, the HBC stated it would consider parting with some land only if the GTR met certain sale conditions. Henry Hulse Berens, a Governor in the HBC, responded to the Duke of Newcastle often. One response was:

> What! Sequester our very tap-root! Take away the fertile lands where our buffalo feed! Let in all kinds of people to squat and settle and frighten away the fur bearing animals they don't hunt and kill! Impossible ... If these gentlemen are so patriotic, why don't they buy us out?[8]

Berens agreed only in principle to surrender a swath of land for a telegraph line. This amounted to approximately 13,000 square miles between the Red River settlement and the Rocky Mountains. The Hudson's Bay and Grand Trunk Railway maintained strong and influential positions, and did not intend to surrender their rights and demands. Both companies too, had English benefactors and origins. Further, the Select Committee of the British House of Commons stated: "It is essential to meet the just and reasonable wishes of Canada to be enabled to annex to her territory such portion of the land in her neighbourhood as may be available to her for the purposes of settlement - with which lands she is willing to open and maintain communications ...."[9] The HBC did begin to "...spar with the British and Canadian governments over legal investigations of its Charter, and on the other [hand] to await [its best] offer."[10] Interestingly, Watkin and others had made several Western land purchases. When Berens learned that Watkin and the government wanted more land than he was willing to partition, Berens closed the door. In a letter of 19 May 1862, "... with regard to a telegraphic communications it is scarcely necessary to point at the prairie fires, the depredation of the natives and the general chapter of accidents as presenting almost insurmountable obstacles ...."[11] Newcastle forwarded a letter in July 1862 to Berens. Newcastle outlined that he would provide "... telegraph service and of securing the means of travelling with regularity to the British territory on the Pacific."[12] Berens agreed only if adequate securities were given for the line's completion. More letters were exchanged, each time the ante increasing. Watkin had "... drawn up ... the establishment of telegraphic communications between the Atlantic and the Pacific through agreements with the Montreal Telegraph Company and the United States Telegraph companies and the construction of a line from Pembina to Fort Gary to Jasper House and Fort Langley."[13] However, in a letter dated 5 March 1863 by the Duke of Newcastle to Watkin, the British imperialists felt that they simply could not "... apply to Parliament to sanction any share in the proposed subsidy ... [and] ... that without a submarine transatlantic telegraph [cable system across the Atlantic Ocean] the proposed line in America will be of comparatively small value to the Imperial Government."[14] Watkin replied on 27 March 1863 suggesting an alternative solution. He wanted the English Government to agree to free land grants provided and in exchange for a "...telegraphic and road communication [route which] passes through any portion of the [Hudson Bay Company] territory."[15] Interestingly, the Montreal Telegraph was not too interested in pioneering the west when they could secure known profits on their current circuits.

Finally, on 15 June 1863, an agreement of 1,500,000 pounds sterling was paid to the Hudson's Bay for Rupert's Land. On 6 July 1863, Watkin went west to establish his plans and to report on "... the possibility of commencing operations for an electric telegraph."[16]

Watkin's telegraph company disclosed their objectives in July 1863. He stated, "With the view of providing the means of telegraphic and postal communications between Canada and British Columbia, across the [Hudson Bay] Company's territory, and thereby of connecting the Atlantic and Pacific Oceans by an exclusively British Route ... one of the first objectives of the company will be to examine the facilities and consider the best means for carrying out this important work; and there can be little doubt that it will be successfully executed...."[17] Watkin and a London financial group even purchased a controlling interest in the HBC. This group felt they could purchase Hudson Bay land then sell it back to begin the Pacific Railway construction under their terms. The plan was not effected.

The Hudson Bay Company responded in a letter 18 February 1864. They felt they would not likely "... receive benefits corresponding to the cost of constructing a line of telegraph ... unless 'the Atlantic and Pacific Transit and Telegraph Company' are prepared to undertake the construction of a road, pari passu [with equal pace], with the telegraph line, the committee cannot ... recommend acceptance."[18] To appease the HBC, Dr John Rae (B: 30 September 1813 / D: 22 July 1893) was placed in charge of the westward route survey crew in 1864. Rae was the perfect choice: he was a surgeon, an expert boatman, swimmer, climber, and, a Hudson Bay employee at Moose Factory. His assigned objectives were to locate the telegraph route, and examine the possibilities of a road, water, and rail route. In his concluding reports to the GTR and HBC, he noted no severe constructional difficulties from Fort Gary until the Yellowhead Pass in British Columbia. However, with Dr Rae was an unnamed GTR official who reported favourably on the railway route, but from St. Cloud on the Mississippi River to Fort Gary, Manitoba and then west. It was this report, not Rae's, which was in direct violation of the "... avowed purpose of establishing communications on British soil."[19]

By 1865 the Western Union Telegraph Company in the United States, had constructed and was operating a profitable telegraphic line on the west coast of the USA into British Columbia. It was an attempt to link Europe with America through Alaska into Russia. But, if Canada could get the railway and telegraphic lines across the Canadian Prairies from Britain to Australia, an All-Red Route for the British Empire would exist.

Watkin's work was complete but it was repudiated. The detailed plans were defeated when the government withdrew their support for guaranteeing part of the shares issued. His dream, goals and reputation were critically injured. Watkin did try to gain the support of the Montreal Telegraph Company under Hugh Allan and the United States Postal Telegraph Company, but unsuccessfully. Ironically, the plan failed because of internal abominations within the GTR. To add to Watkin's distress, on 5 January 1864, the "Victoria Colonist" reported that "Mr. Brydes, the Grand Trunk Manager, has just returned from England with assurances of British Capitalists approval of a telegraphic line, with a railway line to follow."[20] Watkin resigned just as Confederation was about to include Assiniboia and British Columbia. The Grand Trunk lost the western opportunity just as they had lost other interests and routes in the East. The first proposed western telegraph company died on the table.

It is relevant to illustrate one point in Watkin's defense. Watkin has been portrayed in many different sources, as having strictly personal motives for his own profit, and then for the benefit of the GTR. Approximately seven months after his initial proposals, a letter dated 21 November 1862 from the Colonial Under-Secretary to Berens of the Hudson Bay Company shows no such ulterior motives, rather genuine sincerity for the progress of the telegraphic line and railway: "... you must really help me in that Pacific matter. If, on conclusion, you prefer not to be on the board of any company formed to carry out the scheme ... I will endeavour to meet your views ... the Pacific affair is part of our policy - the success of which will make - believe me - a vast difference in the value of your Grand Trunk property ... all I propose to do ... is to get up a [Telegraph] Company, with a capital not exceeding 300,000 [pounds]."[21] In Watkin's Memories, he records that extensive but only tentative surveys were calculated to determine the best telegraphic route. After his trip west, he felt the best route through the Canadian Prairies was to start at Halifax and go on to Riviere-du-Loup (106 miles east of Quebec); follow the GTR to Sarnia, extend the system into Chicago, using the USA Treaty and telegraph circuits from Chicago to St. Paul. From St. Paul, the line would extend back up to Fort Gary then on to Tete Jaune (Yellowhead) Pass.[22] Interestingly, this report was not tabled with the government until Dr Rae proved the above route a year later. Although Watkin's goals were thwarted, he was knighted for his persistence and dedication to Canada's development.

For approximately twenty years, no further telegraphic plans existed for communications across the West. There existed American connections from Winnipeg south to the United States and in British Columbia. But their value was of little economic or political significance. Nonetheless, the Canadian government became more determined to build an all-Canadian railway across Canada in 1867. The United States had completed their nationwide railway. A repeat of American settlement in Canada's west became probable again, as the Northern Pacific Railroad (NPR) was being built close to the 49th parallel. This would allow the NPR to build American railway branchlines into Canada and impede any Canadian railway west. For most American politicians, a Canadian railway through the prairies was undesirable as it would bring strong British financial assistance and immigrants. "The line of the Northern Pacific runs 1500 miles near the British possessions and when built will drain the agricultural products of the rich Saskatchewan and Red River Districts east of the mountains, and the gold country of the Fraser, Thompson and Kootenay Rivers west of the mountains."[23] John A. Macdonald stated "... that the United States government [is] resolved to do all they can short of war to get possession of the western territory, and we must make immediate and vigorous steps to counteract them. One of the first things to be done is to show unmistakenly our resolve to build the Pacific Railway."[24]

Western surveys for the Pacific Railway began in the early 1870s. Sandford Fleming intuitively knew a telegraph system would be vital to the needs of the railway and Canada. However, the Pacific Scandal changed the saga of Canadian telegraphic development. In 1874, the newly elected Alexander MacKenzie replaced John A. Macdonald. MacKenzie was an upright Scot, devoted, cautious, thrifty and extremely honest. MacKenzie's Government, among other proposals, wished to utilize water routes between the Rocky mountains and the eastern terminus on Georgian Bay. Known for his complacency, he delayed this issue whenever he could. In the Spring of 1874, he discarded the Board of Railway Commissioners. This put the Intercolonial Railway under the Department of Public Works - a very slow, and cautious department. MacKenzie would allow for telegraph construction only after the railway line had been determined.

On 19 February 1874, the Canadian Government instructed J.D. Edgar to visit British Columbia and obtain an agreement. This agreement highlighted "... the construction of the road within the time set was an impossibility, and ... [to] make at once a wagon road and line of telegraph along the whole length of the railway in British Columbia and continue the telegraph across the continent."[25] The Government stated it could not maintain the time limit British Columbia demanded. These terms were removed 8 June 1874.

Further, the terms for Confederation were difficult to negotiate. Mr Walkem, a representative from the British Columbia government reviewed the telegraphic situation of the deal. On 15 June 1874, Walkem recorded "... that a wagon road was not desired and it would be useless to the people of British Columbia and that even the telegraph proposed would not be made until the route of the railway was settled."[26] On 23 July 1874, Mr Walkem stated to the Dominion Government again that a railway route must first be determined then the telegraph connection made.[27]

Walkem received an answer from the Earl of Carnarvon, the Canadian Government arbitrator, on 17 September 1874. A basic agreement had been reached concerning the telegraph and railway.[28] Again, the proposed findings included "... the wagon road to be abandoned and a telegraph line postponed until the [rail] road was located."[29] The Dominion Government objected but did realize that building the line, whether railway or telegraph, would appease British Columbia. Official consent and an agreement arrived on 17 November 1874, when the Earl of Carnarvon wrote "... that the wagon road and telegraph line be immediately constructed..."[30] as a combined effort. Interestingly, a surveyed route between Thunder Bay and the Selkirk's was completed in 1875.

In 1877, MacKenzie explained to the House why his government could not undertake building the Pacific Railway. Canada was trapped in its worst economic Depression. Short telegraphic and railway links were constructed when money became available but only as needed. MacKenzie always stressed frugality. Many government projects were either cancelled or placed on an indefinite hold. To his credit, the government did build west from Winnipeg, modestly. A railway and telegraph line was constructed south from Winnipeg to connect tracks and circuits with the St. Paul and Minneapolis. Furthermore, in August 1878, a line was in progress avoiding the worst of Northern Ontario from Fort William to Winnipeg. Under MacKenzie's government the Intercolonial Railway was completed in 1876. A telegraph circuit and wagon road from the Maritimes to Quebec was slowly being finished. However, most telegraphic communications in the Maritimes went through American telegraph companies and connections.

II.
EARLY CANADIAN PACIFIC TELEGRAPH DEVELOPMENT AND THE ALL-RED ROUTE

The Dominion Government was failing miserably with telegraph and railway construction in the west. Their three primary objectives in the early 1870s were to decide where and how to build west; to raise the required money, and, to keep Canadian control while American money and experience were building the poles and tracks.

British Columbia was connected by telegraph to the United States on 18 April 1865. The Collins Overland Telegraph Company in the United States began constructing a telegraph circuit from central California to British Columbia and into Alaska; cross the Bering Strait into Siberia and connect with the Western Union in Europe. P.M. Collins, president of Collins Overland Telegraph, persuaded the Western Union to build the actual line. The Western Union felt this was a plausible route since the Atlantic submarine cable had failed between 1857 and 1866. In British Columbia, the Western Union had a circuit at New Westminster and a cable under the Strait of Georgia linking Victoria and Nanaimo. By August 1866, the Collins Overland Telegraph had reached "... Yale on the east and Hazelton on the north when construction was stopped by news of the successful operation of the Atlantic Cable."[31] The other Western Union line was constructed in 1867 "... through [the] Fraser Valley to Yale then along the Cariboo [sic] Road to Barkerville, a line branching off at Cache Creek to Savona's Ferry and Kamloops."[32] Two years later it was abandoned only to be taken over by the British Columbia government on 7 August 1870; they transferred control to the Dominion Government in 1871 who operated it until the CPR obtained authority in 1881.

The Northwestern Telegraph Company constructed the first operating circuit across the Prairies. This wire stretched from Fort Gary to Pembina with American telegraph connections along the 49th parallel. It was completed by 20 November 1871. Until this time, the Prairies had been circumvented in favour of an American route. Between 1871 and 1874 there was limited progress on the Prairies.

Telegraph contracts west became prolific. Under the Dominion Government, contracts were awarded by section. Their progressive organization with successful contracts included:

Contract 1 - 17 October 1874
Contractor: Sifton, Glass and Fleming
Area: Fort Gary to Fort Pelly

Contract 2 - 30 October 1874
Contractor: Richard Filles
Area: Fort Pelly to Fort Edmonton

Contract 3 - 10 November 1874
Contractor: F.J. Barnard
Area: Fort Edmonton to Cache Creek

Contract 4 - 8 February 1875
Contractor: Oliver, Davidson and Brown
Area: Prince Arthur's Landing to Red River
Completed: Winnipeg, 28 August 1878 [33]

(Thomas Marks, Reeve of Prince Arthur's Landing sent the first telegram to Thomas Scott, Mayor of Winnipeg.)

Contract 5 - 24 January 1879
Area: Winnipeg to Edmonton via Battleford.[34]

Many more contracts were cancelled than were actually constructed. Considering the highly complex organizational skills required for the Pacific Railway, telegraph construction went smoothly. Other telegraph contracts were merged with track laying ones.

Dominion government telegram rates on the Prairies were expensive and service unreliable. In 1878, telegrams between Thunder Bay and Winnipeg cost $2.00 for ten words; Winnipeg to Edmonton was $3.00 for ten words, and $2.00 from Winnipeg to Battleford. Service was sporadic and remarks like 'mail by dog team in a dust storm was faster' represents a common attitude towards the Canadian telegraph on the prairies. Because the technology had not kept pace with the progress demanded, wire failures were common. Also, since there was no mass immigration to the Prairies yet the telegraph appeared extravagant. The government line was unreliable when first installed. Typically: "As the line has been down since Saturday between Hays Lakes and here," read the 6 December 1880 issue of the Edmonton Bulletin newspaper, "we are without telegrams for this issue. A man will leave tomorrow to repair it, and by the next week we hope to be able to give that latest news from the East up to date."[35] This would change under Van Horne.

The Dominion Government selected the Canadian Pacific Railway to build the all-Canadian railway. It was to meet or surpass the agreed terms met with British Columbia. This included the construction and complete takeover of the Dominion Government Telegraph circuits across the prairies. Therefore, the CPR's charter was revised. Although getting a charter has been compared with the ease of getting a hunting licence, it only gave the company specific legal rights. The company had to secure its own success and finances; there were no guarantees. The government could ease the tariff and tax burden but they had their own interests to protect. Therefore, the 15 February 1881 charter, Section 16, gave the CPR exclusive rights:

> The Company may construct, maintain and work a continuous telegraph line and telephone lines throughout and along the whole line of the Canadian Pacific Railway, or any part thereof, and may also construct or acquire by purchase, lease or otherwise, any other line or lines of telegraph connecting with the line so to be constructed along the line of the said railway, and may undertake the transmission of messages for the public by any such line or lines of telegraph or telephone, and collect tolls for so doing .... And they may use any improvements that may hereafter be invented (subject to the rights of patentees) for telegraphing or telephoning and any other means of communication that may be deemed expedient by the Company at any time hereafter....[36]

Two primary functions of the Charter were: to provide telegraphic service to the railway, and to provide consideration for public services. However, many open-ended clauses caused concern for other telegraph companies. For example, "... and may also construct or acquire by purchase, lease or otherwise, any other line or lines of telegraph ..." worried companies like the Great North Western Telegraph if the Canadian Pacific enacted them.

The government would allow for the reduction of tolls charged but "... only after 10 per cent profit had been made on the capital expended in the construction of the railway and only after the net income of the company had exceeded 10 per cent on the same amount."[37] Further, the government permitted the CPR duty free entry of "... steel rails and other track material, of wire, and of all bridge material for the use in the original construction of the railway and telegraph line."[38]

Nevertheless, much conflict developed about the telegraph-railway route. The Dominion Government prepared the original route. When the government turned the work over to the CPR, sections of the line were moved further south towards the 49th Parallel. One change from the original survey went from Selkirk to Humboldt. "It was decided to maintain the rest of the original line with a few changes ... and the Humboldt-Edmonton portion was connected to the C.P.R. in late 1882 ... In 1884, the Department of Public Works decided to change somewhat the government line from Battleford to Edmonton, so that it would pass through ... Pitt, Mooswa, Saddle Lake, Pakan and Fort Saskatchewan."[39] Telegraph maps marking circuit routes are rare to nonexistent at this time. Surveys detailed the railway tracks with the poles understood to follow beside. Nevertheless, the telegraph line preceded the railway and was on the Pacific coast before the rails.

The telegraph station at Humboldt, Saskatchewan in 1885. This telegraph station sent timely news of the Louis Riel's North-West Rebellion. The Union Jack on the roof indicates the 'military' nature of the station. Colonel George Denison is seated in the middle; the other two men are unknown but likely part of the Governor-General's body guard; a third man, also unknown, stands to the right rear of the photograph. Thomas Pike, the operator at Humboldt, stands in the doorway. The telegraph circuit entered the log house at the back. The telegraph pole can be made out between the metal stove pipe and the mud chimney.

(National Archives of Canada, C-753)

In September 1881, a commercial telegraph company tried to acquire the right to build a telegraph line beside all CPR tracks. Van Horne, an experienced telegrapher, felt the offered price was not sufficient to warrant joint use of the right-of-way. Van Horne stated "... as the CPR was putting up its own line and enjoying its own operators as an adjunct to train operations, it should rightfully operate its own commercial service...."[40] The CPR began to initiate steps to form their own commercial telegraph company. By 1882, Canadian Pacific had 181 miles of one-wire telegraph lines and another 714.5 miles of 2- wire lines used for railway business.[41] A circuit operated between Winnipeg and Port Arthur, Selkirk to Pembina and Cross Lake. Of significant interest, no company was contracted to construct an independent telegraph circuit for the railway or for commercial messages. Albeit, independent contracts were awarded to build railway circuits. Attitudes had drastically changed since the Montreal Telegraph began and it was an assumed fact that railways built telegraph circuits and telegraph companies operated the circuits for the railways.

The Great North Western Telegraph Company (GNWTCo) held little hope for the CPR's telegraphic success in September 1883. The GNWTCo had already acquired the Montreal Telegraph and Dominion Telegraph Companies by Agreement and was extremely dominant in the east. A GNWTCo official interviewed by "Electrical World" in New York in September 1883, stated: "I believe the Canadian Pacific is about to try a business for which it is not qualified. They cannot make it pay, they will simply be throwing away piles of money in erecting and running lines exclusively for themselves."[42] The initial reality of this may have been true, however, Van Horne relished the challenge. Within the same month, Canadian Pacific opened their Winnipeg head office on Main Street. From Winnipeg, telegraph circuits were completed to Port Arthur, Medicine Hat and Deloraine. Since the GNWTCo began in Winnipeg and had a circuit to Toronto, Ottawa and Montreal, CPR railway and commercial telegrams were transferred to the competition until they could complete their own line. The CPR had eight operators, one Superintendent of Telegraphs and 897 operating miles of telegraph across the Prairies.

OTTAWA. Con's Patent INFOLD. Canada, Nov. 10th, 1883; U.S., May 27th, 1884. To open, tear off the colored label at the perforated mark.

THE GREAT NORTH WESTERN TELEGRAPH COMPANY OF CANADA.

OPERATING THE LINES OF THE MONTREAL DOMINION AND MANITOBA TELEGRAPH COMPANIES.

This Company transmits and delivers messages only on conditions limiting its liability, which have been assented to by the sender of the following message.
Errors can be guarded against only by repeating a message back to the sending station for comparison, and the Company will not hold itself liable for errors or delays in transmission or delivery of unrepeated messages, beyond the amount of tolls paid thereon, nor in any case where the claim is not presented in writing within sixty days after sending the message.
This is an unrepeated message, and is delivered by request of the sender, under the conditions named above.

H. P. DWIGHT, General Manager. ERASTUS WIMAN, President.

Money orders by telegraph between principal telegraph offices in Canada and the United States.

TELEGRAM.

To Rt Hon Sir J. A. Macdonald KCB

Use this space for Continuation of Lengthy Addresses, or INSTRUCTIONS TO MESSENGER.

Re

No. 3 Check 27 Pd Sumas

Rec'd No.	From	Sent to	Rec'd by	Time
2	N	CD	me	3h

Ottawa, ...188

From Craigellachie Eagle pass BC 7

Thanks to your far seeing policy and unwavering support the Canadian Pacific Railway is completed the last rail was laid this (Saturday) morning at 9:22

W. C. VanHorne

A copy of an original Van Horne telegram. Van Horne's signature is on a Great North Western Telegraph form as the CPR circuit had not been completed from Winnipeg to Ottawa and Toronto. Telegrams were typically transferred to the competition at Winnipeg if going East or south through American connections.

Concurrently, the United States Postal Telegraph Cable Company was being reorganized and the Commercial Cable Company established. Canadian Pacific Telegraphs under C.R. Hosmer, the first General Manager of CP Telegraphs in 1881, developed a working alliance with the two American companies. Since the interchange of freight and commercial telegrams was becoming more lucrative, it proved an excellent opportunity for the CPR. This cooperative established a precedent where the individualistic style of companies was hidden in secret code. The general policy was for Canadian and American telegraph companies to reach legal reciprocal contracts. These agreements would provide the railway with "... free use of the wires for despatching of trains and assistance in the handling of telegrams pertaining to the railway's business ... for its part, the railway provided the services of its operations in the handling of commercial telegrams and granted certain other privileges."[43] This agreement lead to the creation of a North American standardized telegraphic code. The formation of the American Association of Railroads followed. Yet secret coding practices would remain, but now they would not interfere with interchange traffic or operations.

Between 1884 and 1886, Canadian Pacific continued western development. Canadian Pacific's track history is well documented but telegraphic credit is rarely considered. The telegraph was an indispensable instrument in communication and western development. In 1884, Canadian Pacific Telegraphs was through to Laggan, British Columbia and many telegraph and railway branchlines were being developed in all directions. By the end of the year, CP Telegraphs had a net profit of $70,000.

The year 1885 was a turning point. Construction from Port Arthur to Nipigon, Ontario was concluded. Canadian Pacific Telegraphs was now operating in Vancouver. But, had it not been for Louis Riel (B: 22 October 1844 / D: 16 November 1885), the telegraph and railway through Canada's west and Ontario's north would have taken years longer to construct. In the first instance, the Lieutenant-Governor of the North West Territories, the Honourable William McDougall, was ordered "... to take immediate steps for the extension of the telegraph system from the territory to Pembina and for its connexion [sic] at that place with the system of the American Telegraph Company."[44] This included any necessary military action against Riel. McDougall was unable to proceed past Fort Gary and establish the telegraph line because of Riel's attacks. However, on 2 November 1869, Riel was captured and escorted to the American side of the border. In the 1880s, Riel again provided a convenient excuse for the completion of the railway and telegraph. The CPR was also "... turning a profit of $145,000 worth of dots and dashes, thanks mainly to Louis Riel and the newspaper correspondents covering his North West Rebellion."[45] Special telegraph units were used to coordinate and dispatch the movement of the military, supplies and extra trains. One telegrapher was killed, another held prisoner until the Rebellion ended.

Towns were developing alongside the tracks of the Ontario and Quebec Railway between Montreal, Smiths Falls, Peterborough and Toronto. The significance of a wider network of telegraphic communication to these towns began in earnest in 1886. And because they evolved near the CPR, the telegraph would serve their economic needs with little or no inclusion of the GNWTCo. Also, the Postal Telegraph Company, the Baltimore and Ohio Company and the Canadian Pacific Telegraph Company lines were linked. This provided an almost united North American telegraphic service from the big city to the remotest regions. The CPR and Postal Telegraph had a joint line from Vancouver to San Francisco. (Originally the American Pacific Postal Telegraph Company had built a connecting line from New Westminster to southern California. However, when the Postal Telegraph finished their transcontinental line, they absorbed the Pacific Postal Company.) Furthermore, a circuit between Toronto and Niagara Falls, originally operated by the Canada Mutual Telegraph Company, was leased by the CPR.

Bathurst Station, Ontario & Quebec Railway / CPR, 1883. Points of interest: note the first generation CPR telegraph sign at the bay window. Outside, the telegraph pole to the left of the photograph is roofed and not tapered. Four wires come down from this pole to the corner crossarm, across the top of the doorways to a beam of insulators at the bay window to connect with the key, sounder and relay inside. The right pole comes into the back of the station connecting with the telegraph equipment also. Notman took this photograph of my grandfather, grandmother (platform) and great uncle (signal wrench) along the Ontario and Quebec Railway.

(R.G. Burnet)

FIGURE ONE

Principle Telegraph Circuit Route of the Canadian Pacific Railway 1886 (Non-comprehensive)

-Quebec-
Montreal
Dorval
Beaconsville
Vaudreuil
Rigaud

-Ontario-
St. Eugene
Vankleek Hill
Ottawa
Stittville
Carleton
Arnprior
Sand Point
Renfrew
Pembroke
Chalk River
North Bay
Sudbury*
Cartier
Chapleau
White River
Schreiber
Port Arthur
Fort William
Ignace
Kenora
Keewatin
Ingolf

-Manitoba-
Telford
Shelley
Winnipeg
Portage la Prairie
Brandon
Hargrave
Kirkella

-Saskatchewan-
Fleming
Wolseley
Indian head
Qu'Appelle
Balgonie
Regina
Moose Jaw
Morse
Swift Current
Gull Lake
Hatton

-Alberta-
Walsh
Medicine Hat
Brooks
Crowfoot
Calgary
Bearspaw
Canmore
Banff
Massive
Lake Louise
Stephen
Partridge

-British Columbia-
Cathedral
Field
Golden
Moberly
Donald
Rogers
Cut Bank
Rogers Pass
Glacier
Glacier House
Illecillewaet
Twin Butte
Revelstoke
Craigellachie
Notch Hill
Kamloops
Munro
Savona
Spence's Bridge
China Bar
Yale
Vancouver

* from Sudbury the CPR circuit went south to Parry Sound, Bala, MacTier, Medonte, Barrie, Alliston, Toronto and east to Agincourt, Peterborough, Havelock, Tweed, Bathurst, Smiths Falls, Finch, Montreal

In three years, the CPR telegraph system far exceeded the Great North Western's circuits. The Daily Free Press newspaper in Burlington, Vermont noted the CPR's success on Monday, 13 September 1886. The headline read: "Canadian Pacific Telegraph - A New System of Telegraphs to Begin Today - An Important Event". Canadian Pacific Telegraphs had opened Canada's first all Canadian telegraph system to complement their national railway. The CPR "... had four direct wires between Montreal and Winnipeg, two of which will be used for the movements of trains and other railroad matters. This will leave two wires for commercial business."[46] The telegraph also operated from the Atlantic to Pacific, connected with most major USA telegraph companies and trans-Atlantic service to Britain and Europe. Other connections were to the Commercial and French Atlantic Cable Company for trans-Atlantic telegraphic service; from Montreal to Richford, Vermont; Morristown, New York and Buffalo. These connections provided the CPR with access to other American telegraph companies such as the United Lines, American Rapid, Northern Mutual and the Bankers Merchants Telegraph Company.

With the inauguration of an all Canadian telegraphic service, an old dream of Sandford Fleming's was becoming reality. The All-Red Route was first rationalized in 1848. The term was derived from the colour red used by map makers to highlight the British Empire and colonies. Two officers, Major Robert Carmichael Smyth and Lieutenant Millington Synge were the first to describe an All-Red Route. Albeit, their idea involved military protection of the trade route through the Prairies to the Pacific and not a telegraphic communication line. Watkin, in the 1860s, was next to describe an All-Red Route but for a telegraph circuit that would politically and economically unite the fledging British Empire. The united British Empire by telegraphic communication would lay dormant until 1879. Sandford Fleming, Engineer-in-Chief of the CPR and master of the Intercolonial Railway, wanted a telegraph system west connecting the British Empire. Strathcona was the most supportive for this communication route. The vision was achievable: to link Great Britain by telegraph cable and ocean going ships; the landline telegraph and Canadian Pacific Railway across Canada, and submarine cable and ship connecting Australia and New Zealand. The first portion connected with the trans-Atlantic cable at Canso, Nova Scotia, with Great Britain, a distance of approximately 4600 miles, and Vancouver, British Columbia. This was completed on 1 September 1886.

The official opening of the All-Red Route's first segment did not occur until 20 December 1886 at 12:20 pm. A telegram from New Westminster, British Columbia was sent to Old Westminster in England making the first authenticated connection by the Canadian Pacific Trans-Canada Telegraph System (their original but short-lived corporate name). By 1888, they had constructed a line from Montreal through Maine, to connect with Fredericton, New Brunswick and Halifax and New Glasgow, Nova Scotia. The following year, 1889, "... by permission of the Intercolonial Railway System ... a line [was] constructed by [the] Commercial Cable Company for handling of submarine cable traffic for Great Britain and beyond."[47] The second segment was not finished until 1902 when the government owned Pacific Cable Company finished laying the Pacific submarine cable. The true All-Red Route connected the Atlantic submarine cable from Britain and Europe to Canso, Nova Scotia; it then crossed Canada by landline to Vancouver, through another submarine cable under the Pacific Ocean to Australia and New Zealand. Sir Sandford Fleming was given the honour of sending the first telegram on the All-Red Route, 1 November 1902, to all British Commonwealth destinations. CP Telegraphs was uniquely a worldwide system now.

III.
THE FINAL PHASE

Prairie settlement began to mushroom at the end of the 19th Century. Competition for freight, passenger fares and tariffs stiffened between the two other transcontinental railways. Telegraphically, where the Great North Western Telegraph Company held a virtual monopoly in the East, the CPR held one in the West, but the CPR was gaining eastern inroads. The telegraph had become extremely vital by 1896: "... the electric telegraph [had] become a necessity to our political, commercial and social life. It tells ... of the million incidents that make up our national life."[48] As Canadian railways expanded so too did telegraphic train operations and commercial wire services.

In 1899, the CPR completed, "A second wire ... through the Crow's Nest Pass - from Swift Current, Assa., to Nelson, B.C. ... [The line] now gives a first class telegraph service to Fort Steele, Cranbrook, Wardner, Fernie, etc., besides a first-class alternate circuit to Nelson and other Kootenay offices."[49] During this period, there was much reorganization of the company. It demanded streamlined procedures and coordinated technical efficiency. New Divisional offices opened when the railway's expanded their boundaries due to better steam locomotives. Indeed, Canadian Pacific Telegraphs became more aggressive. Canadian Pacific developed solely on their own initiatives by many creative people and top company officials. The competitive telegraph companies could not approach the superiority of Canadian Pacific Telegraphs. Such an example occurred in August 1905. Canadian Pacific "... entered into [an] exclusive telegraph working agreement with the Temiskaming and Northern Ontario Railway commission ... thereby serving the famous Cobalt mining district."[50] CP Telegraphs furnished the needed battery facilities, tariff books and circulars while the Temiskaming and Northern Ontario Commission supplied message and telegraphic forms. Interestingly, the CPR also concluded an intensive agreement with the SOO Line in 1917. This agreement allowed the CPR to construct a circuit along the "... SOO Line from Sault Ste. Marie, Michigan to Emerson, Manitoba via Minneapolis ... [providing] an alternative means of reaching Winnipeg and Western Canada from the East in the event of interruptions on the C.P.R.'s main line in Canada."[51]

Canadian Pacific observed the success the Great North Western News Department was having. Canadian Pacific Telegraphs therefore began leasing a news circuit to the Associated Press in direct competition with the GNWTCo. This circuit, with other railway and commercial messages, could only be used between 8 AM and 6 PM. In 1907 the CPR tried to quadruple the prices charged to three Winnipeg newspapers. The papers objected, so the CPR cut their service, including one newspaper in Nelson, B.C. In 1910, the four newspapers took action against the CPR. The newly established Board of Railway Commissioners of Canada ruled against the CPR in 1911 citing that their 1907 price increase was unfair. The CPR retaliated again by expunging the news revenue service, but would lease their lines to newspapers. By 1911, CP Telegraphs leased a circuit to the Canada Associated Press, an Association of daily and weekly newspapers in Canada. In 1917, Canadian Pacific Telegraphs leased a dedicated circuit to the Canadian Daily Press, which connected the newspapers of Sydney, Nova Scotia to Victoria, British Columbia.

From 1920 on, Canadian Pacific Telegraphs continued to expand. Canadian National Telegraphs was being coordinated and in a few years would become a world-wide telegraph system. Although lines would be constructed on an "as need" basis, the principle construction era was over; more effort would be spent on technological development for faster efficient service. The primary difference between Canadian Pacific and Canadian National Telegraphs within Canada was their philosophical development: the CPT's parent company built railway right-of-ways where communities and farmlands would expand. It followed that telegraph and railway branch lines would be built from the mainline: their development not unlike the Grand Trunk and Montreal Telegraph philosophy. Canadian National Telegraphs grew by legal acquisitions. Also, there was serious talk of amalgamating CNTelegraph with Canadian Pacific Telegraph saving 60 million dollars. (This would surface again in 1947. It was suggested the two commercial departments of each railway company join. This would provide a massive pooling of technological resources to compete against the ever increasing power of The Bell Telephone Company of Canada. In 20 years, the amalgamation philosophy would return.)

During the latter part of the 1920s, the railways embarked on a series of headlining technological advancements and firsts. One achievement was the introduction of the Carrier Current system by Canadian National. Canadian National introduced it 26 April 1927. Within two years the CNR's entire mainline system had been converted. The CPR initiated Carrier Current on 2 October 1928 between Toronto and Montreal.

In 1931, the "... first trans-Canada long distance telephone service was established, [with] the various telephone systems co-ordinating their wire services...."[52] Also, the CPR changed their corporate name from Canadian Pacific Telegraphs "... to the more comprehensive "Communications" as better indicating the complete scope of the Company's activities."[53] It would now be known as Canadian Pacific Communications. Then, in 1932, the CPR secured a joint contract with Canadian National Railways. This contract provided Dominion wide telegraphic communication with the Canadian Radio Commission, later the Canadian Broadcasting Commission. This furnished Canadian National with a unique opportunity to expand their radio network, telegraph and telephone services on their trains.

A telegraphic stock ticker service was introduced by CP Telegraphs in 1937. A ticker connected the financial institutions in Canada and the United States with the Canadian Dow Financial News Service. The ticker service branched into leased wire services with cable, wireless, brokerages and industrial companies. The "Message Transmission Service" provided a series of different commercial telegram rates. According to the time and message length one selected when sending a commercial telegram, the rates varied.

Individual railway circuits still handled most of the commercial and railway traffic. Though there were multiple wires, it was not enough. Slowly, the railway and commercial divisions merged in both

Canadian National and Canadian Pacific. It was found that service remained the same to no detrimental revenue effect.

IV.

CONCLUSION

After World War Two, Canadian Pacific Communications and Canadian National Telegraphs began to solidify their operations. Newer and faster technology was being developed and used. The Telex machine, Deskfacsimilie and the telephone were the three chief opponents dismembering the telegraph. Telegraph significance quickly declined with the conclusion of World War Two. The last official telegraph message in Canada, (except backup service on Vancouver Island October 1973), was sent on 30 May 1972 from Batiscan, Quebec. CP operator Rene Chevalier keyed the final bilingual message to Roland Poliquin in Montreal:

> "Voici Le Dermier message telegraphique in code Morse au Canada. Que fait doni Jesu?"
>
> then in English,
>
> "This is the last telegram via Morse Code in Canada. What hath God wrought?"[54]

CHAPTER THREE

TELEGRAPHIC INVENTIONS

I.
EARLY ELECTRICAL INVENTIONS - 1700 TO 1840

The need to communicate has always been essential. Long ago people went to extremes to pass along news, warn others of danger, help those who were ill, and even gossip. The unassisted human voice is limited and can only travel small distances. To overcome this shortfall, drums once sent stylized secretive codes; smoke signals on hills developed, as did the semaphore flags and Roman signalling towers. But the idea remained the same - to talk with other people over vast distances. Such was the case when Morse developed the telegraph.

Many worked on different telegraphic methodologies. The first known documentation on electric telegraphy is traceable to a detailed article written in a 1753 Scottish science magazine and signed only CM. CM advised the use of insulated electrical wire to protect it from the weather. He also suggested a transmitting code incorporating the alphabet's twenty-six letters, each letter requiring its own wire. At the end of each wire path was a pith ball. As the electrical current was received, the pith ball moved to indicate a letter and eventually formed a message. Also tried were bells that rang when the current was received, and paper strips to record the sent letter. Interestingly, no letter or numeric code was assigned to abbreviations or punctuation; symbols such as the 'comma' or 'period' were transmitted as the full word. Other telegraphic systems were explored by Georges Louis Lesage, Betancourt, Lomond, Reisser, and Francisco Salva, but none of their inventions had a lasting influence.

Claude Chappe created an optical telegraphic mechanism about 1791. Chappe's system had two moveable arms or semaphores attached to a tall wooden post. Stations were built at three to six mile intervals between Toulon and Paris. An observer with a telescope in a similar tower viewed the positions of the semaphore arms, noted the message, then relayed it to the next tower. The same procedure occurred until the message reached its destination. (It was recorded in one source that a canon was fired to alert the next tower in the chain that a message was about to be transmitted.) Transmitted messages reportedly took about forty minutes from Toulon to Paris under ideal weather conditions. By 1844, France had 533 stations spanning 3000 miles. Chappe too, is credited with coining the word "telegraph."

Chappe's mechanism for transmitting code was ingenious. Levers, pulleys and wire were connected to the two position arms. At the top of the tower a vertical post supported a rotating beam or regulator. Each arm or indicator could form seven different positions at a forty-five-degree angle to each other. A combination of forty-nine positions was attainable when combined with the position of the second arm for the twenty-five letters ('J' was not used) and ten numbers. The Chappe system had several disadvantages. One of the most telling problems was the ability for all to see the code being transmitted, especially an enemy; weather conditions like fog, heavy rain or snow could hide or distort the position of the arms, fire was always a danger since the tower was wooden and, night telegrams were impossible. Signal Hill and Telegraph Hill are indications of where the Chappe system was once used.

Claude Chappe's wooden semaphore device. The system used all letters of the alphabet except 'J.' The hut contained the levers and pulleys necessary to change the position of the two arms as indicated in the chart. There was a crew of two to four men stationed at each tower.

(Dots and Dashes - January-February-March 1971, Page 5)

About 1795, George Murray, an English Bishop, developed the shutter semaphore. This was an improvement over Chappe's design with six adjustable shutters located in a box. Murray's code was based on how many shutters were open and closed. Each shutter could move ninety- degrees on a horizontal rod. A total of sixty-four code combinations was possible. Like Chappe's device, Murray's semaphore was located on hills where a definite vantage point was achieved. This device was later adapted to communicate between British ships at sea.

An accidental discovery would enhance and develop electrical transmission of telegraphic signals. The earliest recorded transmission of such an electrical charge was in 1727 and it travelled 880 feet on a type of threadlike wire. In 1777, Alessandro Volta, an Italian physicist, realized that iron wire - common and inexpensive since copper wire had not been fully developed - supported on posts would enhance electrical transmission. But it was Luigi Galvani who discovered that a direct or continuous flow of electrical current along an electrical conductor such as iron wire was sustainable. A direct flow of electricity was further feasible when two dissimilar metals were brought into contact with a moist substance. By 1786 and with refined equipment, Galvani successfully sent electrical current from Como to Milan, Italy. This was the beginning of the voltaic pile or battery used in electrical transmission of a charge. However, Galvani's batteries failed quickly because conductors were not insulated to prevent current leakage.

Between 1805 and 1809, Francisco Salva and Samuel Thomas respectively, introduced an apparatus in which the voltaic pile was used to initiate a signal. At the receiving end the electrical current caused the release of hydrogen in a liquid, however, it could not be determined how to decode the hydrogen bubbles with accuracy. Fourteen years later, William Nicholson and Sir Anthony Carisle lead the discovery of enhanced electrical current. They found that electrical current passing through specific liquids caused them to break down into their basic elements.

Further electrical concepts and theories developed around 1819. Hans Christian Oersted (also Orsted), a physics professor at the University of Copenhagen, studied magnets, batteries and electricity. Oersteds' voltaic pile experiments lead to the basic elements of battery composition: cells of copper, zinc and diluted acid. Oersted demonstrated that a magnetized needle could be deflected when an electrical charge was transmitted along wire. The stronger the signal, the greater the needle deflected. Using Oersted's work, J.S.C. Schweigger found that a needle's deflection could be increased by surrounding it with several turns of wire. Schweigger's work lead William Sturgeon to produce the first electromagnet about 1825. Sturgeon increased the intensity of the magnetic field by introducing soft iron cores into the coils carrying the electrical current. These cores were a horseshoe shape that effectively attracted the iron bar armature.

In 1820, Andre Marie Ampere applied electricity to an electromagnet. Ampere created the earliest version of a needle telegraph around 1829. By using a battery bank with higher output, he could move and deflect needles more effectively.

Joseph Henry expanded Sturgeon and Oersted's electromagnet theory. By 1829, Henry (1797-1878) had formulated his theoretical laws governing coil design for electromagnets and applied voltages for maximum electrical impact. Improving on Oersted's work, Henry (occasionally referred to as John) conducted experiments using silk insulated copper wire; by increasing the windings inserted inside the coil, Henry developed an electromagnet capable of lifting 750 pounds. Next, he experimented with a battery bank of twenty-five cells. These cells generated enough current to cause an electromagnet one mile away to move and ring a small bell. In 1831, he published a descriptive magnetic theory proving that an electromagnetic machine could lift several thousand pounds. Henry also produced a crudely designed electromagnetic telegraph to explain his theoretical ideas. Joseph Henry "...had discovered the electro-magnetic force which aided Morse in his discovery of how to apply that theory to an up-and-down movement of a metallic mechanism."[1]

With Henry's work, increased electrical charges produced successful electromagnet applications. Henry enlarged the scope of many theoretical applications by Galvani, Nicholson, Carisle and Oersted, and proved their usefulness to an electrical telegraph. Later John Henry would meet Morse and they would discuss the possibilities of an electromagnetic telegraph.

Concurrently, two Englishmen experimenting with another form of telegraph were William Fathergill Cooke (1806-79) and Charles Wheatstone (1802-75). Improving on Ampere's original visual telegraph, the 1837 Wheatstone and Cooke telegraph allowed one to actually read the transmitted letters on a display board. This patented visual telegraph had six wires with five magnetized needles. When an electrical current was sent through the wire, various combinations of letters could be created. The sending operator directed a needle to the required character on his hand turned magneto-generator board. When activated, this pointer transmitted a predetermined number of electrical impulses. At the receiving terminal these impulses drove a similar needle and when they stopped moving, the needles pointed to a letter or number. Some letters required as many as three complete movements of a single needle. The stopped positions were noted on paper and a message was created. In total, twenty letters were arranged on the dial. Thirty letters could be sent in about one minute, taking another few minutes to decipher and form the syntax.

The Wheatstone-Cooke needle telegraph was tested on 25 July 1837 between Euston and Camden Town. In 1838 the Great Western Railway (GWR) tested it between Paddington (London) and West Drayton. By 1839, the GWR used this telegraph system for train operations in transmitting signal information only. Bells, used by the dispatcher to alert a station of an incoming message, rang in every station on the entire circuit. Futile experiments were conducted to create selective ringing and this would not occur for another fifty years. The Wheatstone-Cooke telegraph proved its communication value with the capture of a murderer. John Tawell, spotted boarding a train at Slough after committing a brutal murder, travelled to Paddington. His every move was reported by railway telegraph operators to the police. Once into Paddington Station, he was arrested. The needle telegraph of Wheatstone and Cooke had been the key player in his arrest. Interestingly, the Great Western Railway dispensed with this telegraphic system about 1842.

In September 1894, a similar telegraphic chase occurred in New Brunswick, Canada. A brutal convict escaped from the Dorchester Penitentiary. He ran to the Intercolonial Railway yard where he had been a fireman. From the yard he stole a yard steam engine. "When the crew noticed her pulling out fast, the train dispatcher keyed the entire line wanting an 'open-line' for the 80-mile division - orders [were] then issued to a crew to chase but all that was available was a seventy-two-inch diameter 'goat' ... [but] because of hot bearings and drawing boxes, [they] had to stop to cool by throwing water on them ...."[2] All train movements of the convict were reported, even his depleting coal supply which eventually caused his locomotive to stall. Regrettably, the convict managed to escape.

II.
THE GROUP OF FOUR

Born 27 April 1791 in Charlestown, Massachusetts, Samuel Finley Breese Morse was the eldest of three brothers. Samuel's father, Jedidiah was a clergyman in one of Boston's Calvinistic churches and a noted geographer. His mother was Elizabeth Ann Breese. Samuel always had an interest in art, specifically painting. His talent was so obvious he attended Yale University in 1805, graduating 1810. To develop his artistic skills, he studied in London and Paris. When the United States Capital Building was being constructed, Morse applied to do interior paintings. The President, however, rejected Morse supposedly over a trivial personal matter. This disapproval caused Morse to virtually abandon painting as a career. (Morse died 2 April 1872 and is buried in Greenwood Cemetery, Brooklyn, New York.)

Returning to the USA from Europe aboard the packet ship "Sully" after studying Art at England's Royal Academy, Morse began to question transmitting electrical current along wire. He developed an awkward electromagnet model. Alfred Vail and his brother George agreed to build an improved model of Morse's telegraph. The father, Judge Stephen Vail, provided Speedwell Village Iron Works Factory in Morristown, New Jersey for one-quarter interest in the profits of the telegraph patent.

While developing his telegraphic device, Morse became a Professor at New York University in 1835. Although he taught primarily Art, Morse developed many telegraphic designs and constructed various telegraphic models on his own time. Leonard Gale, a colleague and professor of Geology and

Mineralogy, discussed with Morse the tremendous potential of electrical principles in 1836. It was shown that the electrical circuit would not operate past a 20-mile limit without radical changes to the equipment and increased voltage. As a result, Gale invented the relay that improved electrical current strength and its corresponding send/receive signal over a greater distance. Gale also showed Morse could increase voltage by increasing the number of wires around an electromagnet. This resulted in several relay stations being established to boost the electrical signal for Morse's experiment.

Morse's first functional model was completed in 1836. The instrument consisted of a crude picture frame secured to a desk top. A metallic punch on an armature was activated by an electrical current, causing the punch to drop onto a thin paper strip. The thin paper strip was pulled by a hanging weight over a desk top edge. In January 1838, Morse and Vail sent and received their first telegraph along two miles of wire strung throughout the factory. The message keyed to Morse was suggested by Vail's father and read: "A patient waiter is no loser."[3] The second test, repeated for several university friends, included Leonard Gale. "The demonstration consisted of stringing 1700 feet of wire about the walls of the university."[4]

Interestingly, the United States Congress originally considered the construction of a Chappe visual telegraphic semaphore system between New York and New Orleans in 1837. At this time, "telegraph" meant a system consisting of semaphores, flags and long wooden poles. The first Chappe system in the United States was between Martha's Vineyard and Boston in 1800. The New York-New Orleans telegraph would have constructed telegraph stations at least eight miles apart. Circulars were sent by the Secretary of the Treasury to military personnel and scientists requesting their analysis of this proposed system. Samuel Morse replied, stating the Chappe system was ineffective at night or in bad weather. Morse further stressed that secrecy would become nonexistent during war and the tower would be an easy target to destroy. In February 1838, Morse substantiated his electrical telegraph to the United States House Committee of Congress. Congress was then left to debate and decide the merits of the Chappe's semaphore system and Morse's electrical telegraph.

Meanwhile, Morse established a partnership in February 1838. The four partners and their telegraphic ownership were: Samuel Morse (56.25%), Alfred Vail (12.5%), F.O.J. "Fog" Smith (25%), and Leonard D. Gale (6.25%). On 13 January 1843, Gale resigned from the partnership due to ill health. Erza Cornell became a partner, obtaining Gale's 6.25% later that year. It should be noted that the Group of Four agreed that any further telegraphic inventions or any such improvements by any of the partners became their joint property and liability. Interestingly, Professor John Henry was offered a partnership but he declined.

Morse and "Fog" Smith travelled to England to apply for the patent. In England they demonstrated the electromagnetic recording telegraph to Wheatstone and Cooke, nonetheless, they dismissed it as not commercially profitable. The English government also turned Morse down, citing a published article in a scientific magazine that described the results of Morse's electromagnetic telegraph but did not describe its method of operation. Disgusted, Morse and Smith travelled to France. The French granted a patent on 29 August 1839. The American patent was not awarded until ten years later after their first refusal. However the United States government made it retroactive to 1839 when they realized the significance of the telegraph to American business and security interests.

On 23 February 1843, Congressman Smith, who was the House Committee Chairman on Commerce, introduced a Bill "...appropriating $30,000 for [the] building of a test line between Baltimore and Washington [some 40 miles apart] ... The line was along the right-of-way of the Baltimore and Ohio RailRoad."[5] John H. Latrobe and L. McLane also made the completion of the first telegraph line possible with many Resolutions. Because he was known as a good promoter with a strong political position, Smith secured the Bill. Secretly, Smith did not reveal his real interest to Congress as a financial partner in the telegraph invention.

It was beside the Baltimore & Ohio RailRoad that Ezra Cornell would construct Morse's first telegraph circuit. Cornell first attempted an underground cable in 1841. He purchased the patent on a special plough that would dig a trench, lay and bury the cable in one operation. Baltimore and

Ohio RailRoad foremen were given the task to cast a ploughshare to Cornell's specifications. The plough would make a furrow two inches wide and twenty inches deep. Wire would be fed into sixty foot sections of lead pipe. The pipe rolled off the plough's platform by pulleys. A plumber, armed with a firepot and soldering irons, followed behind joining the pipes. The plough, pulled by sixteen oxen, could accomplish about one-half to one mile per day. However, it was not until after ten miles the connections were tested. This test revealed an electrical short in the circuit. As a result the entire furrow was dug up, the wires inspected and it was learned that ground made a better return wire for the entire circuit.

Morse ordered all work halted. Workers immediately assumed the construction was a failure and began demanding full pay for the entire forty miles or they would sue. Fischer, the original contractor, was left to resolve this problem while Cornell and Morse dealt with the electrical short. They removed the wire from the lead pipe covering and redivided the four iron wire strands of the double current into separate wires. An insulation mixture of cotton saturated in gum shellac was applied. This insulation failed as the short continued. Out of desperation, Cornell suggested the wire be striped of all insulation and strung bare in tree branches acting as temporary poles. Broken glass bottle necks served as the first insulators. When this series circuit was again tested, it operated successfully. The law suit then ended and the men went back to work.

On 24 May 1844, the US Supreme Court sanctioned Morse's telegraph system. Annie Ellsworth, the daughter of H.E. Ellsworth, Commissioner of Patents, selected the first official telegram for transmission. Her message came from The Bible, Numbers 23:23, "What hath God Wrought!" and signed "SFB Morse." The telegram was then transmitted from the Court building by Morse to Vail who waited in the Baltimore and Ohio Railway's Mount Claire station. The telegraph in North America was now initiated. Morse then approached the Washington Government wishing to sell the telegraph to them, however, they would not purchase the invention for the suggested $100,000. Greatly angered, Morse began his own company, the Magnetic Telegraph Company that followed the stagecoach road between Washington, New York and Philadelphia. (Much controversy has arisen about the first telegram's appearance. Today, it often appears with a question mark at the end, however, the exclamation point is Biblically and historically correct).

The telegraph for railway operation had a tumultuous beginning in North America. The first application was used as a signalling mechanism and railway business messages along the B & O. Later, on 22 September 1851 Charles Minot, a Superintendent on the New York and Erie Railroad used the telegraph to dispatch his train. Minot's train was held in a siding waiting for a late train to pass. Minot's conductor, W.H. Stewart, learned from the commercial telegraph operator that the ruling train had not reached Goshen thirteen miles away. A ruling train in the 1850s had the right of one hour against an opposing train of the same class. The Conductor on Minot's train was going to wait out the hour before making any further decisions. Minot, anxious to get to Port Jervis, stormed into the telegraph office. He sent a wire to the Goshen agent to hold the ruling train since it had not passed him. Boarding the locomotive, Minot ordered locomotive engineer Isaac Lewis to proceed, but Lewis refused, as such an order was contrary to accepted procedures. Minot then cracked opened the throttle. On arrival at Goshen, Minot found the ruling train had not arrived. Again, he telegraphed ahead to see if the ruling train had left Middletown. Since it too had not arrived, a new wire order was issued to the agent at Middletown to hold the ruling train until Minot arrived. In this hopscotch-like maneuvering, Minot's train travelled to Port Jervis where the westbound was finally met. Minot proved the advantage of the telegraph for safely speeding up train operations. Later, Minot established the first detailed telegraphic train order rulebook that became the foundation of train operations in North America. Interestingly, his grave marker in Harriman, New York, "... commemorates [the] world's first telegraphic train order, 'Run To Goshen Regardless of Opposing Trains' ...."[6]

By 1861, telegraph wires spanned the continental USA from Missouri to San Francisco becoming the first transcontinental telegraph line and replaced the fabled Pony Express on 24 October 1861. It is significant to note that the United States military exploited the telegraph for improved military communications. During the American Civil War the Union Army made effective use of mobile telegraph units.

In Canada, as previously discussed, telegraph wires parallelled the Lake Ontario-St. Lawrence River waterways along the Lake Shore Road. Later, telegraph circuits went through Sarnia and Windsor to connect with various American telegraph companies. Canadian messages were then transferred and retransmitted across the continental USA to California. From California, the message went north to British Columbia circumventing the Prairies until the Canadian Pacific Railway joined British Columbia with the rest of Canada above the 49th Parallel. Later, Louis Riel unwittingly helped to establish the Canadian telegraph in the Prairies and north of Lake Superior in particular because of the Metis Rebellions.

In short, Morse invented the telegraph key and greatly influenced the development of the Code. A total of sixty-two people claim to have invented a telegraph, but Morse made the first practical application of Henry's electromagnetic theory.

CHAPTER FOUR

TOOLS OF THE TELEGRAPH

INTRODUCTION

The successful telegraph operator in 1884 was one who would be "... thoroughly familiar with all practical features of the apparatus and mechanisms of the telegraph."[1] Jessie H. Bunnell, (1843-1899) the most known and accepted telegraph key and sounder manufacturer, went on to insist that, "It is not only a great aid to the prospects of advancement for the operator to have a thorough knowledge of the electrical theory of telegraphing and to understand all about the batteries, the wires and the instruments, but this is for him an opening page to that great book of wonders, Electrical Science...."[2] Like musicians, telegraph operators were fine tuned to many different aspects and styles involved in getting the job done. The tools and skills of the telegraph operator were extensive and complex.

I.
THE KEY

The telegraph key is the sending instrument. It is used to momentarily break the electrical circuit to transmit the dot and dash sequence that form a message. The function of the key is like taking two live wires, briefly touching them together and creating an electrical spark or break.

Morse's original arm and key were made of brass with some iron parts. It had a vertical travel adjuster for changing the distance between contact points that created the electrical spark. This key also pivoted on a central arm. The one disadvantage was the electrical shock the operator sometimes received on the finger tips, as no insulation existed between any of the metallic parts. If a current had too much charge, the operator would certainly feel the power of electricity through the key.

Many other early manufacturers created some interesting designs, but nothing practical developed until Bunnell established a company dedicated to the manufacture of his telegraphic design in 1878. The Bunnell key revolutionized, standardized, and replaced the cumbersome Morse instrument.

Bunnell's key was simply constructed, easy to repair and adjust. When operating the key, one pressed down on a knob of hard rubber or ebony, and this acted as insulation between finger tips and metal. Platina was used in contact areas under the lever or hammer and connected with another platina point on the frame or anvil. Platina was used since it rarely fused, rusted or tarnished. A platinum wire was attached to both points where contact would be made between the hammer and anvil. The wire itself was usually no more than an eighth-of-an-inch in diameter. The lower point was surrounded with rubber acting as insulation and preventing an electrical connection between the hammer and anvil. When the knob was pressed down, electrical contact was made and a tiny spark created.

Attached to the Bunnell key frame were two, three, or no legs. Each leg served as a terminal line to connect the electrical circuit. Telegraph keys with legs were generally stationary instruments often found in stations, terminal centres and commercial offices. A legless key developed a few years later, was semi-portable. Secured to a small wooden platform, a legless key could be moved around a small office or desk like today's computer mouse. This style was much preferred by dispatchers and operators.

An arched lever opened the circuit for transmission. It was known as the circuit-closer. Operators were instructed to always close the key once business was completed as an open key drained the batteries. If this occurred, managers and supervisors took action so it was not repeated; demerit points, fines or dismissal could occur.

(a)

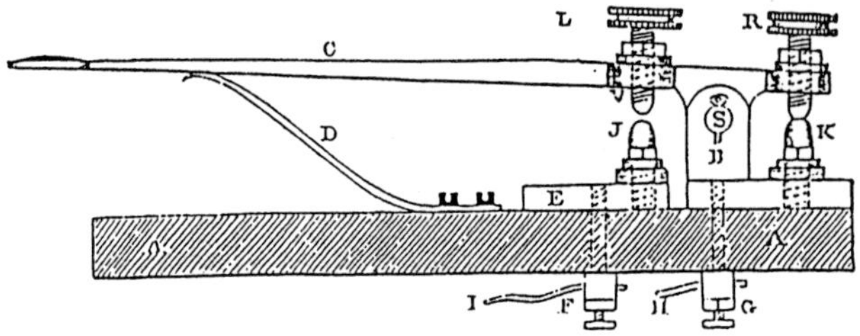

(b)

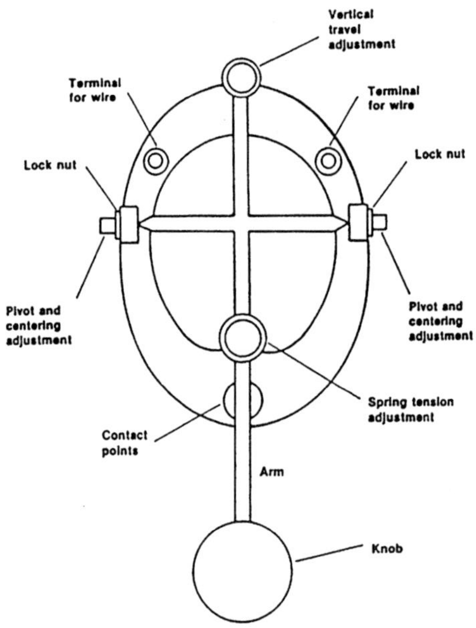

(c)

(a) Sending key from Alfred Vail's 1845 book.
(Burnet, R.G., Canadian Rail #425 / F. Angus Collection)

(b) Diagram detailing the various parts of a standard Bunnell key.
(Source unknown)

(c) A standardized Bunnell key used by my grandfather, uncle and mother at Port Union, Ontario during the 1920s.

The Bunnell key required the operator to use up and down finger pressure and wrist movement to operate the key. By making fine adjustments with the adjusting screw the hammer and anvil could be brought closer or further apart. This allowed the operator to alter the tension of the lever and spring to suit his transmitting speed and skill. When pressure was released, the spring automatically raised the key preventing electrical contact. Easy adjustments of the platnum points were made by turning the pivot adjusting screws.

Interestingly, after several years of sending with a Bunnell key, some operators experienced the ‘glass arm’ syndrome. The malady consisted of their arm ‘locking up’ into an uncontrollable painful cramp. Doctors felt that "... paralysis of the arm [was] due to continued use of certain muscles in the wrist when sending."[3] Doctors also suggested it could be a nervous complaint. The considered remedies included complete rest for the arm; wrap the arm in a cold pack to relieve the pain of his ‘glass arm’; learn to send with the other hand or, buy your own typewriter since railway and commercial telegraph companies did not supply them until about 1900. It was a sign of the times that if one did not work one did not eat, so operators often learned to send with the other hand. The typewriter would later provide an alternate method for receiving.

A deviation from the Bunnell key was the Autoplex. Horace G. Martin developed this style in 1903. In 1904 Martin introduced the Vibroflex or "bug" which became an extremely popular alternative to the Bunnell key. The vibroflex required a swaying left-right motion of the arm, wrist and fingers. With a vibroflex, operating speed was greatly enhanced and paralysis rarely occurred. Many styles developed however, the Bunnell key was the standard most railway companies supplied for their operators.

It is noteworthy that stations were distinguished by the presence or lack of a telegraph. Small stations had only a station agent or station agent/operator at the key while small or rural stations maybe no one at all. A station however was elevated in status if it had its own telegraph operator who was kept separate from the station master or agent. The operator often received the privileged location in the station's tower or second floor where he could closely eye the tracks and community. Note, if a station had no operator or key, there was most likely a commercial telegraph office in town run by the main telegraph companies.

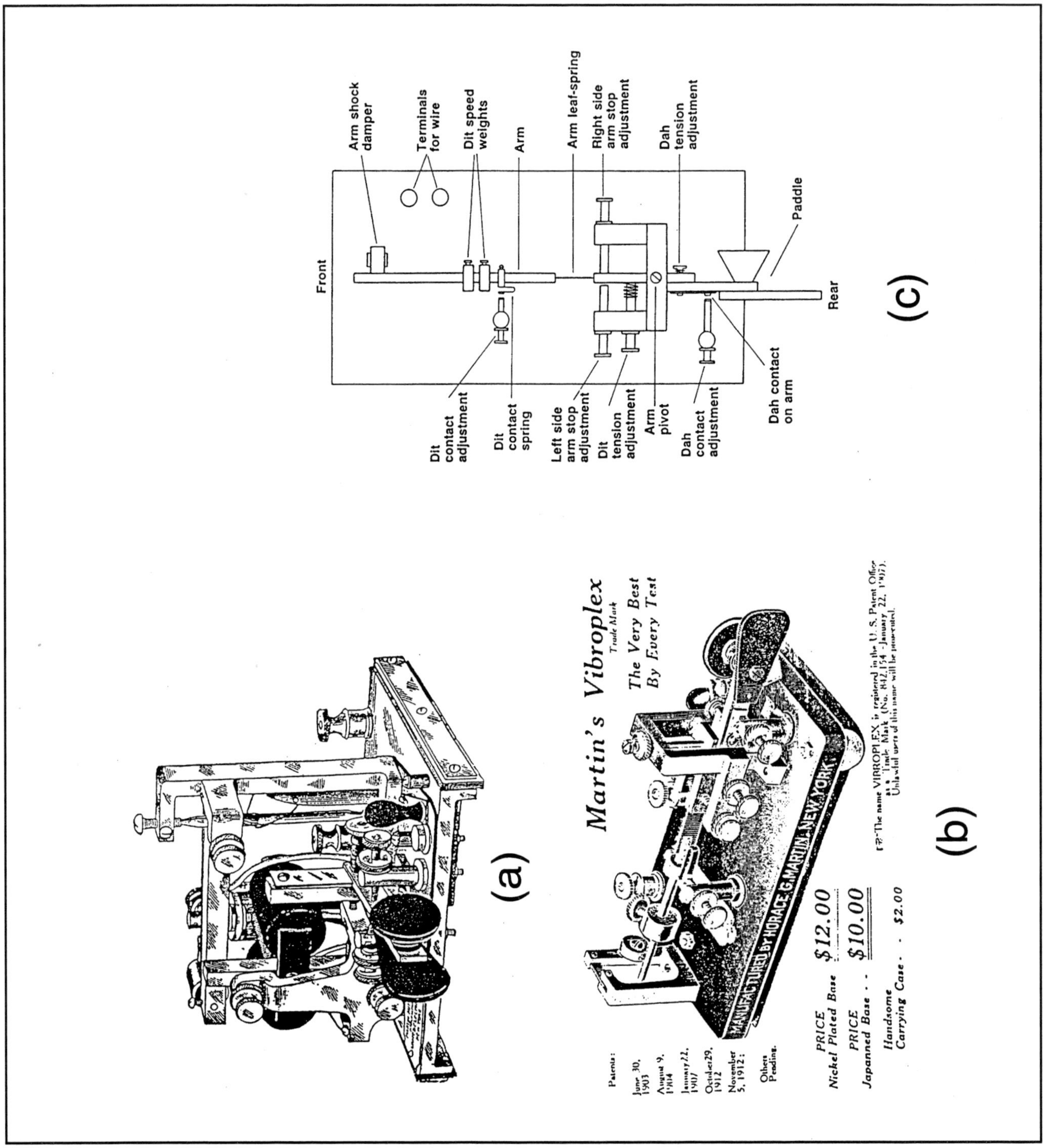

(a) An Autoplex ad from Sear Catalogue 1900. The Autoplex was not commonly used.

(b) Martin Vibroflex or 'bug' cal915 from a Sears Catalogue.

(c) Diagram detailing the various parts of a standard Martin Yibroflex.
(Source Unknown)

II.
REGISTER

The register was the original Morse receiving unit. It had a crank attached to a weight on a cord. This was wound up and tightened so it could rotate the rollers to draw the tape through the imprinting device. The electrical dit, dah and space signal was encoded onto paper tape using a single stylus pen for each electromagnet. A two-pen register also existed. A stamped "V" was recorded on the tape. By counting the "V"s and making note of their relationship and spacing to one another, it was possible to translate the "V" back into a numeral or letter and in turn a word. Operators then read and translated the paper tape impressions into a textual message. The register too, had a clockwork mechanism that continually forced paper tape under the stylus.

The register had many operational problems. Some common technical problems included: paper tape jams were frequent requiring the sending operator to retransmit; tape was destroyed by the stylus punching holes in the tape if unattended or unadjusted, and, if the tape ran out there was often no extra supply of tape. Other disadvantages included: the breaking of the weighted cord; a constant winding required for the clockwork mechanism; the operator could easily make a decoding mistake of the imprinted V's. Errors intensified when the mechanism jammed, tearing the paper tape. It was rare to find a register in use after about 1865.

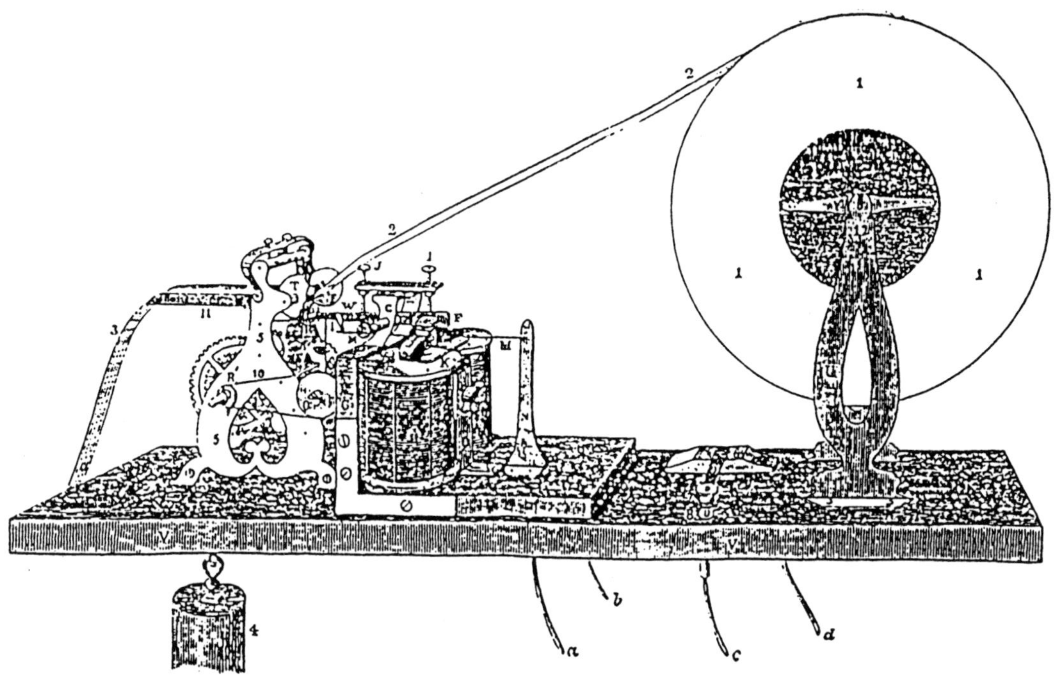

An illustration of a paper tape register from Alfred Vail's 1845 book. Note the similarities of this register to the sounder photograph. By removing the tape pulling mechanism and adjusting the armature position and shape, the basic sounder is present.

(Burnet, R.G., Canadian Rail, Issue #425 / F. Angus Collection)

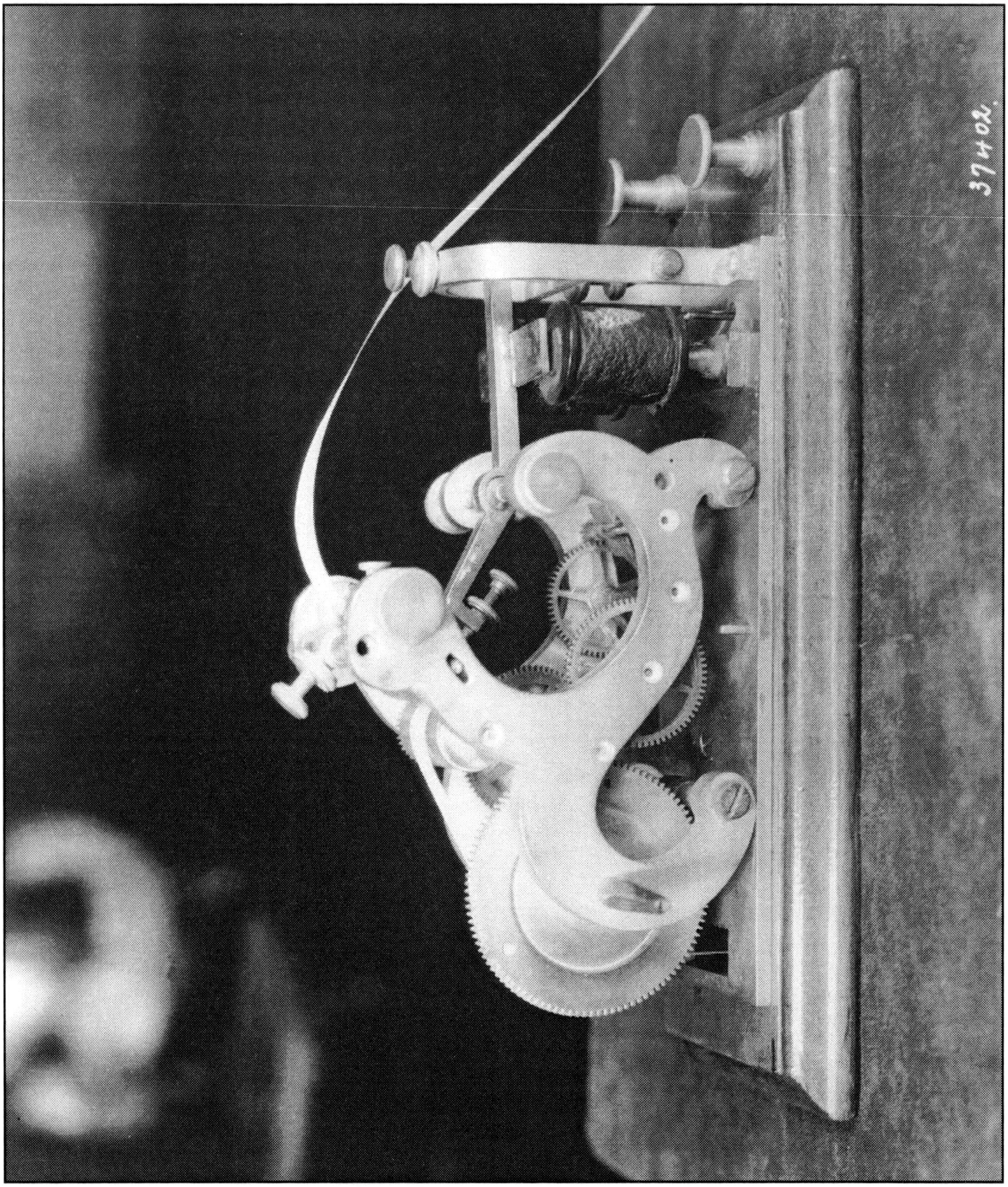

An example of an original 1850 Morse receiving register. This style was first used by the Toronto, Hamilton, Niagara, St. Catharines and Niagara Company and Montreal Telegraph Company until about 1860. The Morse sounder did not appear until 1856.
(CNR Public Affairs, #37402)

III.
SOUNDER AND RESONATOR

The sounder, invented by Morse about 1856, was first discovered by a teenage operator. Originally, the receiving operator decoded imprinted tape impressions from the register. The impact of the stylus striking the paper tape made distinctive rhythmical sound patterns. Jimmy Leonard, a young fifteen year old, and his friend Joseph Fisher, could receive Morse faster than anyone. The boys simply listened to the stylus pen tap dit-dah patterns while striking the register's paper tape. When this feat was reported to the area railway superintendent in 1854 or 1855, Jimmy was praised for his skill. Jimmy demonstrated his ability to Morse who was amazed at the accuracy and speed of fifty-five words per minute. Morse soon made alterations to the original paper register and the first sounder appeared about 1856. By removing the paper weight and feed, changing the stylus to an iron armature and altering its position, a sounder was born. Variations of this device developed for about fifteen years until Bunnell patented his sounder. It is interesting to note that an American law forbade sound reading until 1858. In Canada, the Toronto Hamilton Niagara and St. Catharines used a Morse register. It can only be assumed that sound reading was not practiced until after 1860.

Not until 15 February 1875 did J.H. Bunnell revealed his patented sounder. The sounder is an electromagnet. Specifically, two bars of soft iron with rounded heads of hard rubber are on top. The round bars in the spools are called cores. A moveable flat piece of iron or armature on top and in front, is attracted by magnetism to the cores, or withdrawn by the spring when there is no core magnetism. The armature is suspended above the magnetic spools with a style or pin. Each of the two spools, mounted vertically, is wound with about an inch-and-a-half of copper wire to about five ohms resistance. When electricity is sent through the spool wires, it magnetizes the iron cores. The actual power of the electrical force depends entirely on the battery's power output. When the magnet is magnetized, the lever moves down and makes a tapping sound. The armature is forced back up by a spring. It was recommended that no more than one-sixteenth-of- an-inch be allowed for the armatures movement, however, it could be adjusted with the adjusting screw.

Standard Bunnell Sounder used by my grandfather, uncle and mother at Port Union, Ontario during the 1920s, and Prince Albert cigarette tin. This size of cigarette tin was preferred by operators as it easily slipped between the sounder and resonator resulting in an amplified sound.
(R.G. Burnet)

Two types of sounders were used: the mainline sounder and the local sounder. The mainline sounder was connected to the dispatcher's mainline circuit. It could adjust quickly to different circuit electrical conditions and operating speeds. Because of greater distances from a terminal to station(s), Direct Current (DC) resistance could range from a low 30 ohms to as high as 240 ohms. The local sounder operated on a separate relay. This was generally connected to its own battery source. Its DC resistance ranged from four to four-hundred ohms.

The resonator was used to amplify the sounder's sound. One style was the standard resonator or butterfly. This resonator had adjustable wings on each end and when both were open it appeared like a butterfly. It was supported by a back piece slanting upwards. The resonator could be found either mounted in a fixed position or on a two-or three-piece adjustable arm. A second style was the pedestal resonator that was stationary and rested on a pedestal. Resonators were generally solid mahogany or cherry wood. Because the sounder did not have any amplification of its own - other than the tapping of the armature and a resonator box - in later years, operators amplified this sound by placing an empty shirt pocket size Prince Albert tobacco tin behind the sounder.

IV.
RELAY

The relay was an essential mainline component in the telegraph system. Until this device was invented, transmitted signals had a limited range. As a circuit went further away from its terminal source and power, the signal weakened. The relay allowed the electromagnet to repeat, enhance, or increase the circuit's electrical capacity when it was not strong enough to carry it. The vertically mounted armature, hung in front of horizontally laid electromagnetic poles and coils. This position allowed for very rapid vibration when electricity passed through it. The vibrations were automatic, based on two physical forces: one, a retractile spring and, two, the magnetic force generated by the electromagnet. When no magnetism occurred, the spring pushed the armature away from the electromagnet. Direct Current resistance ranged from 35 to 150 ohms. Wire gauge and armature size used by the relay was typically smaller. Bad weather could reduce a signal's strength; only accepted adjustment regulations, issued by the railways, could be practiced.

CANADIAN NATIONAL RAILWAYS

COMMUNICATIONS DEPARTMENT.

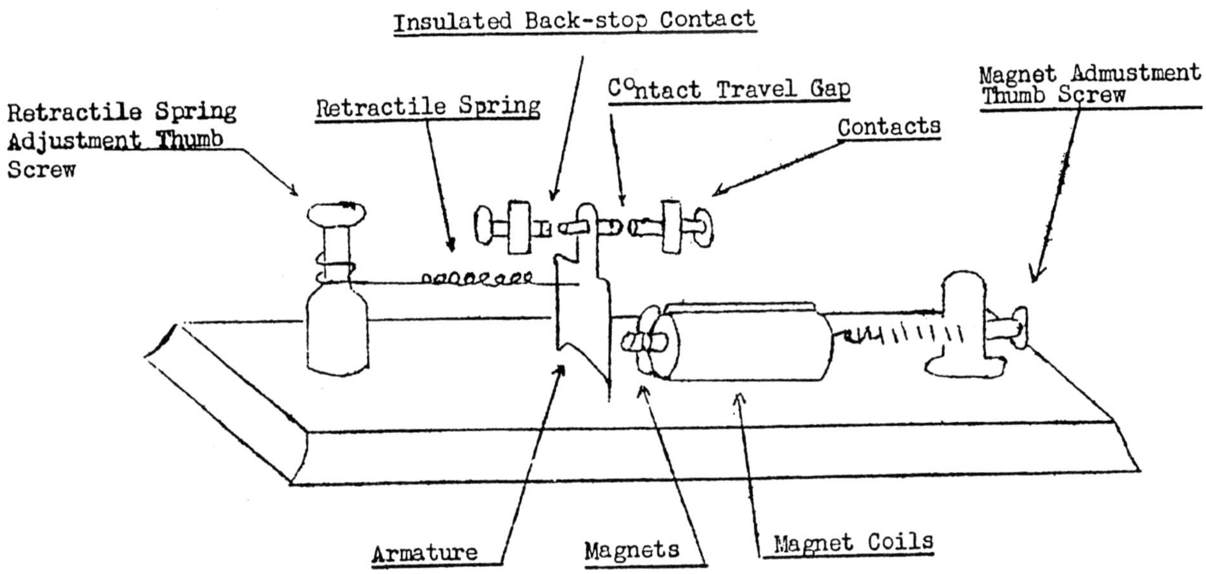

H. C. Milne,
Superintendent.

Canadian National Railways advice for maintaining relays. This supplement was issued by the CNR Communication Department ca1940s.
(Courtesy G.W. Horner)

AGENTS, OPERATORS - CANADIAN NATIONAL RAILWAYS

During inclement weather, (rain, fog, drizzle, wet snow, sleet,etc.) the margin of adjustment of a Way Wire Morse Relay is decreased. To get the Relay to respond to Line Signals, adjustments should always be made by decreasing the magnet pull through the relay by withdrawing the magnets from the armature. A Morse Relay has to be re-adjusted for every weather change.

THE CORREST METHOD OF ADJUSTING A MORSE RELAY, is:

1. Retractile spring

-with your key open
-tension just sufficient to return armature to back-stop
-close your key
-IMPORTANT: tension ALWAYS JUST SUFFICIENT to return armature to back-stop

2. Adjust magnet coils

-with your key closed
-turn magnet adjusting thumbscrew to clockwise direction
-until YOUR SOUNDER goes open

3. Re-adjust magnet coils

-with your key closed
-turn magnet adjusting thumbscrew in anti-clockwise direction
— until YOUR SOUNDER just closes
-with this adjustment most stations should be readable
-SLIGHT adjustment may be required to bring in more distant stations
-with this adjustment YOUR OWN SENDING may not be clear at times

4. Abnormal weather conditions

-with key closed
—turn magnet adjusting thumbscrew (clockwise)
-full extent of travel may NOT be sufficient to open your sounder
-then, AND ONLY THEN, should retractile spring tension be increased
-until YOUR SOUNDER just opens

5. Contact travel gap

-not more than 10 Mils., or
-(.010", which is about thickness of message blank doubled)
-contacts always clean and flat-faced.

G.Gottfred, Superintendent,
Communications Department.

Canadian National Railways advice for maintaining relays. This supplement was issued by the CNR Communication Department ca1940s.
(Courtesy G.W. Horner)

Though the designs differed, relays were used extensively to boost the electrical signal. Weather conditions did effect the ohmic loss. This photo shows a standard Bunnell relay with a spring contact breaker repeater. It was part of my grandfather's own telegraph set from GTR Darlington Station.
(R.G. Burnet)

A composite photograph detailing various early telegraphic equipment used by the Grand Trunk Railway of Canada.

(CNR Public Affairs, #47513)

THE Northern Electric AND MANUFACTURING CO. LIMITED

TELEPHONE EQUIPMENT

FOR STEAM AND ELECTRIC RAILWAYS

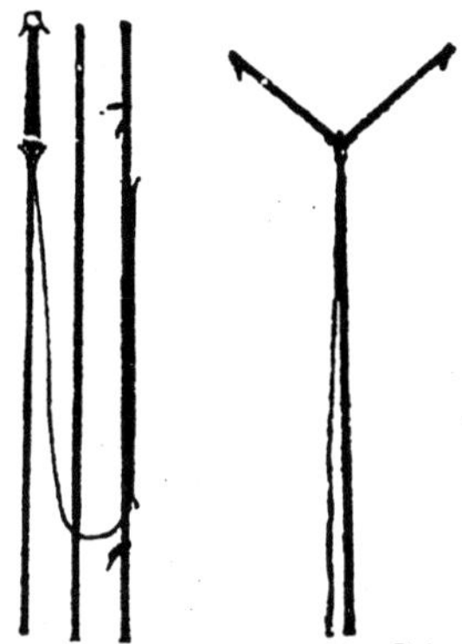

No. 3 Line Pole Disjointed

No. 3 Line Pole Assembled

LINE POLES

Made of hickory and are adapted for use in connection with metallic circuits; the spreaders are of sufficient length to engage wires spaced a distance of 2 feet apart. Connectors are provided with a cleaning device. Equipped with 100 feet of 2-conductor cord.

Used with Nos. 1330-E, 1331-E, 1332-E, N3-A, and 1332-E portable telephones.

Trade No.	Mfrs. No.	Price Each
089495	3	$7.84

No. 4 Line Pole

Similar to No. 3, but arranged for connecting the portable telephone to the line wire of a grounded circuit. Equipped with 10 feet of single-conductor cord.

Used with No. 1314-A portable telephone.

Trade No.	Mfrs. No.	Price Each
089496	4	$7.23

RICHARDSON LINE POLES

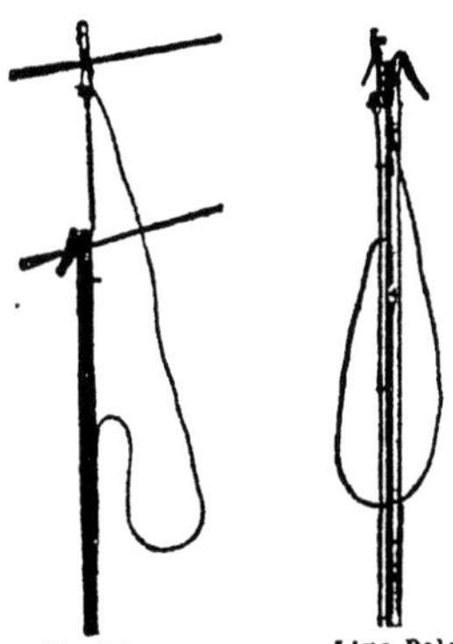

No. N2

Line Pole

Arranged for use in connection with metallic circuits. Can be connected to line wires in either horizontal or vertical planes which are spaced any distance up to 5½ feet. The adjustment of one-wire clamp is controlled by a cord. The length of the cord connecting the line to the telephone set is 100 feet.

Used with Nos. 1330-E, 1331-E, 1332-A, 1332-E and N3-A portable telephones.

Trade No.	Mfrs. No.	Price Each
089497	N2	$8.78

Line pole examples sold by Northern Electric. Although these pole styles were for telephone use, they were fundamentally the same style used by conductors to connect their portable sets.
(Courtesy Bell Canada, Document No. 10304-1, 1913 and 21110, 1920, Northern Electric Catalogue)

Portable telegraph set used by my grandfather at GTR Darlington Station. Early agent's were expected to repair the lines in their section when troubles occurred. However, this practice ended by 1910 with full-time maintenance workers, although 'old-timers' were still known to venture out and do their own maintenance. Wires were connected to the terminals of the set (one missing here) from the wires overhead. A line pole might be used to hook directly to the circuit or by a brakeman climbing the pole. Portables were used at wrecks, by superintendents, found in business or private cars, repair crews. On the front is stamped "W.U. Tel. Co., 150 ohms"; on the backed is stamped "Prepared by H. Thau."
(R.G. Burnet)

V.
PORTABLE AND PRACTICE SETS

Portable sets were carried in the van (caboose) or on passenger trains. In case of wrecks, track damage, needed clearance, and generally to be in touch with the dispatcher, a portable set was used by the conductor. Most sets were self-contained: a key, sounder, possibly a relay, the necessary two to four poles to hook the wires on one board, and a few dry cell batteries. When a message was required, the tailend brakeman generally climbed a telegraph pole dragging twenty or thirty feet of copper insulated wire. He then hooked or clipped the ends to the dispatcher's designated circuit on the telegraph pole diagrammed in the Employee Timetable. The conductor then attached the other end of the wires to the proper terminal poles on the portable set and keyed his identification, news or requests. Line poles of ten to twenty-three foot extendable heights were also used to hook onto the wires from the ground. However, their use depended on the conductor, as line poles were very awkward to use.

Practice sets were made for children, but skilled and learning operators also purchased them. They were not unlike the portable sets required on trains. Children were fascinated by the operation of the telegraph, especially its rhythmical music. These sets were 19th- early 20th century educational toys. They came complete with a dry cell battery, wire, key, sounder, manual and codebook enclosed in a wooden carrying case. Practice sets did not generally have the terminal poles to connect to actual circuits. They could be purchased through mail order for $2.00 or $3.00 in 1900.

40 J. H. BUNNELL & Co.,

THE "MORSE" LEARNERS' OUTFIT.

Price, **MORSE OUTFIT COMPLETE**, with Battery, Book of Instruction, Wire, Chemicals, and all necessary materials for operating...... $3 75
"Morse" Instrument alone, without battery...... 3 00
Cell of battery complete...................... 65
"Morse" Learners' Instrument, without battery, sent by mail, prepaid....................... 3 50
Morse Instrument, wound with fine wire, 20 ohms resistance, for use *only* on outdoor lines of from 200 feet to 10 or 15 miles in length, price, without battery, &c................... 3 75
Sent by mail, prepaid.......................... 4 00

Battery cannot be sent by mail.

The above will be sent C. O. D. to all points if one-third of the amount of the bill is sent with the order.

Remit by Draft, Money Order, Registered Letter, or Postal Note.

112 LIBERTY STREET, NEW YORK. 41

THE MORSE OUTFIT

shown on the opposite page, is a full-size, well-made, complete

Morse Telegraph Apparatus

of the latest and best form for learners, including handsome Giant Sounder and Curved Key and a large Cell of the best Gravity Battery, latest form.

It is the best working set of Learners' Instruments for short or long lines, from a few feet up to 20 miles in length, yet produced.

See the next pages for information in regard to the cost of Instruments, Wires, Batteries and Materials needed in putting up

SHORT LINES OF TELEGRAPH.

A "Morse" Learners' outfit as described and advertised in the "Student's Manual for the Practical Instruction of Learners of Telegraphy" by J.H. Bunnell, 1884, Page 40-41
(Reprint of original Catalogue)

TYPEWRITERS—ALL MAKES.	
Remington No. 6	$25.00 to $55.00
Smith Premier No. 2	$25.00 to $55.00
L. C. Smith No. 1	$45.00 to $65.00
Remington No. 10	$65.00 to $75.00
Monarch No. 1	$65.00 to $75.00

¶ Remit $5.00, state kind of typewriter desired and your machine will go forward immediately. Give certificate number when ordering. You can pay the remainder at rate of $4.00 per month.

¶ If you need a new cylinder on your typewriter, send the old cylinder in to me with $1.25 and we will furnish you a new one—you can put it on. Ames & Filstead Platens. Best made.

Agate Pointed Styluses	50c.
Hudson Word Counters	$3.00
Phillips Code	$1.00
Martin Night Letter Counter	$2.00

Remit by Money Order or Check to

J. E. ALBRIGHT, 253 Broadway, New York City

Sole Selling Agent for Martin Vibroplex Company.

Remington Typewriter Ad from Sears Catalogue 1898 and Price List from ca1915.

Note the Underwood typewriter in the foreground and the cutdown centre of the desk. This was a standard design for typewriters found in stations and terminal areas. Note the relays, fixed Bunnell keys, sounder and resonator on a swinging pedestal so two operator's could use it. The man in the front row left is wearing the green eyeshades so typical of old agents, operators and dispatchers, but no one is wearing the arm bands. "Messages intended for transmission are hung in front of the operator on hooks supported by a brass standard; after being sent over the wire the messages are filed on a hook on the standard resonator, & collected every half hour...." (Railway and Shipping World, March 1899, Page 93)
(CNR Public Affairs, #53873.7)

VI.
TYPEWRITER

The earliest appearing typewriter was in 1714 by Henry Mill. The first practical model was not patented until 1868 by Christopher Sholes (1819-1890). Later, Remington (1816-1889) patented his namesake on a greatly improved typewriter in 1876. Underwood was another very familiar typewriter manufacturer used by the railways.

Another advantage for the typewriter was its ability to make legible and multiple copies of train orders. At first, train orders were individually handwritten in ink only. A lead pencil was never allowed because an error could be erased. The tissue paper or flimsy was used to avoid correcting mistakes as it tore easily. An order with an error or tear was voided and reissued. If multiple copies were required, they too had to be written out by hand. If a double or triple header was scheduled, multiple copies of the same order had to be written and individually hooped to each locomotive. Canadian Pacific, in some selected areas, allowed agents to use indelible pencil. The order could then be put in a large tissue page ledger. A wet piece of cotton was placed over the tissue and taken to a press. The press caused the indelible pencil to leave an exact copy on the underlying tissue. Later, carbon paper was developed to circumvent this practice. With the typewriter and carbon paper, multiple copies could be clearly typed in one setting with fewer delays, more efficiency, and legible. It was at this point the railways began supplying "mills" to the busiest stations first.

Originally, typewriters "... had to be purchased by the operators and staff if they wished to use them ... [they] cost approximately $150.00...."[4] Typewriters found in stations, telegram offices and wire services used only capital letter type - lower case was for everything else like railway correspondence, reports and so on. Interestingly, with the introduction of the vibroflex, many operators purchased the cheaper bug. When they were bumped to another station or office, it was easier to carry a bug than a cumbersome and heavy typewriter.

VII.
BAUDOT'S MULTIPLEXING

Many alternative methods to the key developed. Increasing the speed of a message was sought and the typewriter provided the source for many other creative inventions.

Emile Baudot, a French telegrapher, developed multiplexing. In 1872, he reasoned that several messages could simultaneously be sent over a single wire and be automatically printed on a roll of paper. In essence, Baudot's multiplexing system used synchronized sending and receiving rotary dials at each terminal. A five-unit code was used for the printing receiver. The five levers on the dial were set in alternate positions by electromagnets. The magnets were then synchronized with little notches on a disk. A typehead continuously rotated and this was attached to a printing hammer - its appearance not unlike a daisy wheel printer/typewriter today. As each code was discerned by the receiving unit, the signal was printed on paper tape timed precisely to the hammer's movement. However, because the device required sophisticated timing and precise mechanical parts, it was not exclusively adopted. Baudot's multiplexing was ahead of its technological time. There were too many technical and mechanical problems that could not be solved since the needed technology did not exist. But, Baudot did foreshadow the bases for later inventions.

VIII.
THE PRINTING TELEGRAPH, MORKUM PRINTER AND THE START-STOP PRINTER

In 1846, Royal E. House patented his printing telegraph. The basis for this invention was a stepping typewheel. Transmission was via a piano-like keyboard. It is recorded that up to three thousand words of press releases could be sent in one hour. Paper tape was used to record the transmission. The House system operated between New York and Philadelphia around 1848. The next year connections were made to Boston, Buffalo and Chicago.

The Printing Telegraph

Printing Telegraph. Note the pianolike keyboard for 'playing' in the message for transmission and the circular roll. Some sources claim that this could be the beginning of the early gramophone. (Source unknown)

Although his idea was unique, House's printing telegraph was not functional and failed. There were several reasons for its failure. These included: its sensitivity to weather conditions; its construction was extremely complicated and costly to repair, and, its sending speed could not appropriately be maintained by the receiving unit. Since the receiving unit feeder rolls were not synchronized with the sending unit, the typewheel over-typed other letters creating unreadable messages.

D.E. Hughes of Kentucky, in 1855, produced a similar printing telegraph to R.E. House. Hughes' device was highly technical and mechanized. Synchronization was the key to its clockwork mechanism. His machine offered a continuously rotating typewheel with a piano-like keyboard for information. An electrical current was emitted at a point in the receiving unit's typewheel revolution to strike the chosen character. This caused a paper strip to be brought in contact with the wheel and the character was printed. Its success was short lived.

G.M. Phelps of Troy, New York, created a variation in 1859 that became known as a combination set. Phelp's used the best ideas from House and Hughes. Phelps used a step-by-step principle. The typewheel contained the letters of the alphabet, numerals and special characters evenly spaced on tymes on a circular plate. Letters were advanced by a ratchet. One character therefore had its own electrical impulse to cause it to strike the paper when selected. With this device, operators could attain sixty to one-hundred words per minute. However, it did not develop and suffered with similar problems and House's Printing telegraph.

Phelp's hardware was based on House's original ideas and Baudot's multiplexing. Interestingly, Phelp's combination set became the precursor to the telex. In August 1912, Canadian Pacific installed a Morkum Printer between Montreal and Toronto. Tests were successful and it was extended to main line areas in Halifax, Winnipeg, Ottawa, Moose Jaw and Vancouver in 1913. However, there were many problems with timing and its use was limited.

By 1920, the Morkum printer was accurately synchronized. E.E. Kleinschmidt, the Morkum Company in the United States, and Siemens-Halskelo in Germany, collaborated to develop teleprinters or teletypewriters. England was involved to a lesser extent. The Morkum Printer or MUX consisted of a typewriter-like keyboard organization. It had a printing mechanism that could output the message as a continuous line of single type or in page format from a roll of paper attached to the back of the stand alone unit. An electric motor was used to control the speed and this was again synchronized with clutches and coupling devices. The MUX made use of Baudot's five unit code idea. When a key was depressed, five bars took one of two unique positions representing one character or numeral. A clutch then automatically engaged allowing transmission to occur. The receiving unit worked in exact synchronization with the transmitter. It could transmit sixty to one-hundred words per minute.

Another device was the Start-Stop Printer. This machine was suited for simple, one-channel, stop and start printing of telegraphic messages. It was widely used between 1924 and 1938. The messages were typed electronically on to tape. The tape was manufactured with adhesive on the back and passed through a wick moistener as the message progressed. An operator wore a cutting thimble on the index finger. As the tape exited the printer, it was separated and affixed to a blank telegram form. Interestingly, it was the first machine to also allow for typographical corrections on paper at the sender's machine. Canadian National and Canadian Pacific used this printer, but not on a wide scale.

An example of a teleprinter producing the punched tape in the MUX department, 10th Floor, National Building, 347 Bay Street, Toronto. Telegraph tape had five large holes with the one smaller centre feeder holes used to pull the tape through the machine. The five holes correspond to the five-unit code being sent. Cable coded tape would have two or three rows of large dots with the feeder holes in the centre. Interestingly this style of tape was used by UNIVAC and IBM computers during the 1950s and 1960s in computer data processing.

(CNR Public Affairs #52061-1)

IX.
THE BATTERY AND MOTOR GENERATOR

Two types of batteries were commonly used: dry cells, and the gravity or wet cell. The dry cell battery was primarily used on an open circuit system. Practice and portable sets generally had a dry cell battery for power. It was preferred since it was portable, clean and did not spill corrosive acid. Since the dry cell was disconnected after use, it retained most of its charge for a considerable length of time. However, the dry cell could not be reused or recharged when void.

The battery had significantly improved by 1870. As previously noted, the battery developed from a stage of low electrical impulse to one where a bank of 25-cells could be connected in parallel or series lifting a 750-pound load. It too, was better chemically. However, power drained quickly and was finicky. J.H. Bunnell stressed proper battery management for optimum electrical effect for his equipment. In a 1884 published manual by Bunnell, early batteries had to be setup according to the following instructions:

> Fill the jar about two-thirds full of water. Place the copper foot [or crows' foot] in the bottom of the jar in such a way that the leaves of the copper are spread out like an 'X' with the copper wire extending upwards out of the jar.
>
> Next, carefully drop into the bottom of the jar about one- half pound of blue vitrol (copper sulphate) and two ounces of sulphate of zinc.
>
> Hang the zinc in the jar ... and the battery is ready for operation although it will not work at its best power until it has been in use about three days.
>
> The battery should be kept supplied with enough sulphate of copper so that a blue color can always be seen in the liquid at the bottom of the jar, rising to within an inch of the lower surface of the suspended zinc.
>
> (The above procedure applied to this battery, often termed a 'gravity' battery. The term 'gravity battery' developed from keeping the sulphate of copper and sulphate of zinc apart by the force of gravity.)

There were other precautions. Agents had to be careful a battery did not freeze as this impeded electrical output. Furthermore, a glass jar size of five-by-six inches was recommended. Batteries were best stored in a high cupboard, near a chimney or wood stove so dampness could not occur. The heat, too, kept the battery from freezing during the winter and helped the battery to operate more efficiently by promoting the needed chemical reaction. Batteries had to be easily accessible for renewal of chemical materials and inspection.

The electrical action of the gravity battery depended on the strength of the chemical fluid and its connections. A hydrometer was employed to determine the gravity (density) of the two fluids. When inserted into the solution, the hydrometer needed to register between fifteen and thirty-five degrees; twenty-five degrees was considered a 'charged' battery. If less than fifteen degrees was registered, not enough zinc was in the solution, making it a weak electrical charge and the battery was short circuited. If over thirty-five degrees was registered by the hydrometer, too much zinc was present and must be withdrawn by syringe; this was an overcharge that could damage relays and other telegraph equipment.

Batteries were connected in series or a parallel circuit. For a series circuit connection, located in the centre of the cell was the positive carbon (or copper) pole; it was attached to the outside negative zinc pole of the next cell. In this way, the voltage of the series cells added and combined to increase the amount of the circuit voltage. Series circuits were the most common and efficient electrical circuits. A parallel circuit had limited use. The negative poles were connected as were the positive poles. The resultant voltage was like one cell.

Northern Electric Company
LIMITED 17

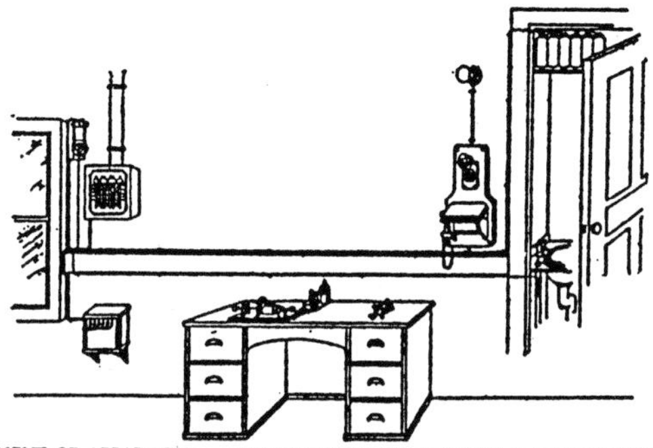

ARRANGEMENT OF APPARATUS TELEGRAPH AND TERMINAL TELEPHONE STATION
RAILWAY COMPOSITE SYSTEM
Fig. 20

ARRANGEMENT OF APPARATUS INTERMEDIATE TELEGRAPH AND TELEPHONE STATION
RAILWAY COMPOSITE SYSTEM
Fig. 21

Battery placement in stations. Top: Storage batteries are on the top shelf of the cupboard. Bottom: Batteries are under the bay window desk, which is not an ideal location as generally this was a damp and cold freezing area in many station structures.
(courtesy Bell Canada, Montreal, Document 3853, Page 17, undated)

Operators were reminded to always close the key after a transmission was completed. An open key drained power. In terminal areas, some dispatchers elected to have a foot switch. The foot switch was used to " ... close the transmitter battery in place of the key operated by hand. This is used to prevent waste of battery and the introduction of noise on the line ...."[5] (It was also used with early telephone technology.) Furthermore, operators were warned to always disconnect their instruments from the battery circuit when leaving the station for the night. Those not complying with these fundamentals needed a good excuse; dismissals, fines and demerits were known to occur. Interestingly, the exhausted battery liquid was sometimes used to clean the station floor of grime, soot and dirt. The spent acid was swept across the wooden floor planks. The weakened chemical would react with the floor boards lifting the stains and dirt. A clear water wash followed. The battery's zinc electrode was removed and scraped clean, and the copper electrode was washed of dirt and sediment. Both were reused as long as possible.

Batteries were inexpensive. For example, in the 1890s a four to ten cell battery cost 65 cents; ten pounds of Blue Vitrol was $1.00 and two pounds of sulphate of zinc, 20 cents. Also required was one pound of office wire for 50 cents; one box of steel staples for twelve cents; approximately 200 to 800 feet of Number 12 galvanized iron telegraph wire at 50 cents per 100 feet, and, one-20 ohm Morse instrument for $3.75.

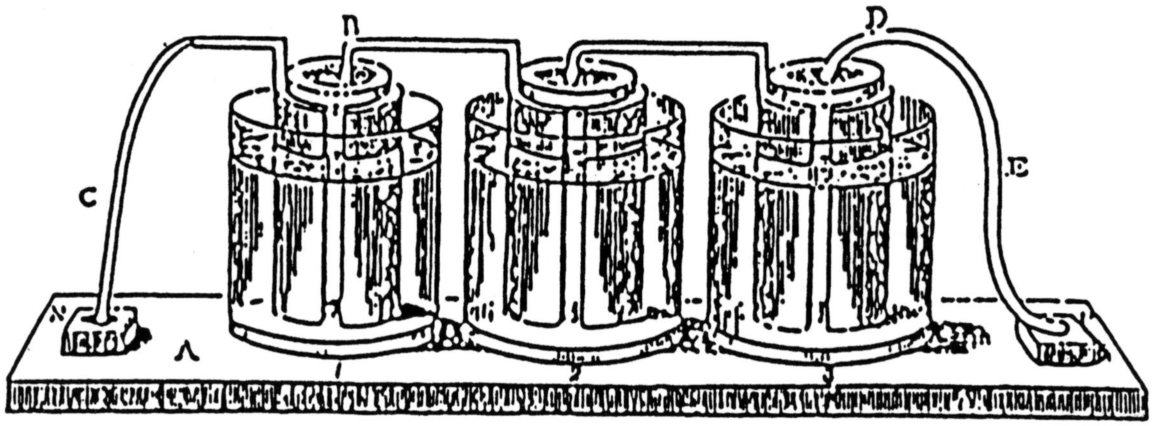

Battery setup from Alfred Vail's 1845 book .
(Burnet, R.G., Canadian Rail #425 / F. Angus Collection)

THE WESTERN UNION TELEGRAPH COMPANY

PLANT DEPARTMENT

RULES FOR THE CARE OF IRON GRAVITY BATTERY

TO SET UP A CELL:

1. CLEAN EACH JAR THOROUGHLY.
2. PLACE COPPER IN JAR, WITH LEAVES SPREAD OUT.
3. DROP BLUESTONE CRYSTALS DOWN ONE SIDE OF JAR, AND FILL 3 SECTIONS OF THE COPPER, AS IN FIG. 1. ABOUT 1 1/2 LBS. OF BLUESTONE IS ENOUGH. PUT NO BLUESTONE AT ALL IN THE OTHER 3 SECTIONS.
4. HANG IRON IN POSITION, AS SHOWN, AND THEN POUR OVER IRON A QUART OF CLEAN TOP SOLUTION SAVED FROM OLD IRON OR ZINC BATTERY, ADDING ENOUGH WATER TO COVER IRON 1 INCH.

NOTE: IF NO OLD TOP SOLUTION CAN BE OBTAINED, SHORT-CIRCUIT CELL FOR AS LONG AS NECESSARY; OR IN CASES OF EXTREME EMERGENCY USE STARTING SOLUTION OF SMALL HANDFUL OF EPSOM SALTS, DISSOLVED IN QUART OF WATER.

5. DO NOT PUT ANY OIL ON IRON CELLS.

FIG. 1

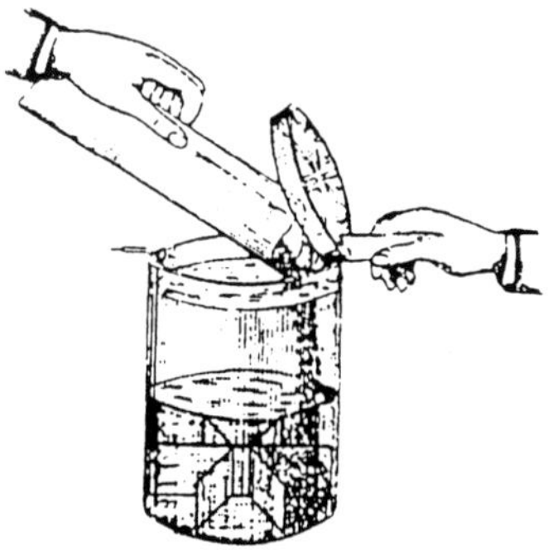

FIG. 2
Method of adding fresh charge of bluestone without removing any connecting wires. Tilt the iron up, resting it on the top of the jar, and drop 1 1/2 lbs. of Bluestone Crystals into the proper 3 sections.

MAINTENANCE:

1. CELLS, CABINETS AND RACKS MUST BE KEPT THOROUGHLY CLEAN AND DRY, AND ALL CONNECTIONS CLEAN, DRY AND TIGHT.
2. ABOUT EVERY THREE WEEKS, REMOVE 2 INCHES OF THE TOP SOLUTION WITH AS LITTLE DISTURBANCE AS POSSIBLE, AND REPLACE WITH WATER. THIS SHOULD BE DONE TO KEEP THE CELLS IN PRIME CONDITION AND AVOID CREEPING OF SALTS.
3. THE TOP OF THE SOLUTION MUST ALWAYS BE ABOVE THE IRON. REPLENISH WITH WATER AS NECESSARY.
4. WHEN THE FIRST CHARGE OF BLUESTONE CRYSTALS IS NEARLY ALL DISSOLVED, ADD A SECOND CHARGE OF 1 1/2 LBS. IN THE MANNER SHOWN IN FIG. 2 WITHOUT DISMANTLING THE CELL.
5. WHEN THE SECOND CHARGE IS USED UP, CLEAN OUT THE CELL AND SET IT UP ANEW.

TO CLEAN A CELL:

(a) DRAW OFF AND SAVE THE CLEAN PORTION OF THE TOP SOLUTION (IRON SULPHATE).

(b) CLEAN THE JAR AND WASH THE IRON. RUB THE WHEEL OF THE IRON WITH A DAMP RAG IF NECESSARY, BUT KEEP THE TERMINAL CLEAN AND DRY.

(c) SET UP THE CELL AGAIN AS ABOVE, USING THE RETAINED CLEAR SOLUTION TO START THE ACTION OF THE CELL.

(d) USE THE SAME IRON OVER AND OVER AGAIN AS LONG AS ANY PART OF THE WHEEL REMAINS. USE THE OLD COPPER OVER AGAIN IF IN GOOD CONDITION. USE UP ANY GOOD UNDISSOLVED BLUESTONE.

BATTERY REFUSE:

SAVE ALL OLD UNUSABLE COPPERS AND COPPER DEPOSIT AND SOFT COPPER MUD. PLACE THEM IN A CONTAINER AND ASK DISTRICT PLANT SUPERINTENDENT OR SUPERINTENDENT OF TELEGRAPH TO SECURE AND ADVISE SHIPPING INSTRUCTIONS. OLD ZINC RECOVERED FROM ZINC CELLS CONVERTERD TO IRON SHOULD ALSO BE SPEARATELY SAVED AND DISPOSED OF IN THE SAME MANNER.

THE LUGS OF OLD USED IRONS SHOULD BE THROWN AWAY.

G.M. YORKE,
VICE-PRESIDENT, IN CHARGE OF PLANT.

NEW YORK, AUG. 26, 1916

Rules for the Care of Iron Gravity batteries. Note the Western Union date of August 26, 1916 and many interesting 'routines' that the agent had to perform.

Motor generators gradually replaced the Bluestone Gravity batteries used throughout most telegraph systems. Canadian Pacific developed the following practice around 1896:

> ... whenever more than one source of power could be obtained, it was replaced by motor generators and battery as backups ... where only one source of outside power can be obtained, storage batteries are used to bridge over any breakdowns in power ... [and] where more than one source of power is available, motor generators are used ....[6]

Canadian Pacific's Ottawa terminal is believed to be the first Canadian telegraph company to use electrical motor generators with battery backup in 1896. After 1930, electrically run generators were used to supply power to the telegraph system along major mainline routes with storage batteries on standby to bridge power failures. Branchline traffic generally relied on battery power until about the 1940s.

X.
TELEGRAPH WIRE AND INSULATORS

Bare wires had been used as line conductors as early as 1816 by Francois Ronalds. But, when the wire was buried the circuit failed. Ronalds then developed a protective insulation mixture of glass and pitch to cover the wires. But, the insulation rotted away quickly exposing the circuit to electrical shorts. Wheatstone and Cooke experimented with buried telegraph wires. Again, water seeped into their underground pipes and shorted the circuit. Other insulations tried were: tarred cotton fibre, silk, ivory, bone, hard and soft rubber, dry wood and gutta-percha.

A complex science of manufacturing telegraph wires evolved. After repeated failures with buried cable, success was found by stringing individual wire on poles with glass insulators. The best method was open or bare wire construction. The wire used by Canada's first telegraph company, the Toronto, Hamilton, Niagara and St. Catharines, used galvanized uninsulated iron wire. It had a useful life expectancy of about fifteen to twenty years. Some sections on this circuit may also have used bronze, another fair conductor.

Iron wire, readily available, easy to manufacture and very durable, had a history of electrical failure. Most common failures resulted in moisture impeding the circuit. Because iron wire was tense when strung from pole to pole, it often snapped in strong winds, heavy rains or with the build up of snow and ice. Canadian Pacific forty years later, used galvanized iron wire in #9 or #6 gauge. The CPR in 1886 felt that #6 gauge was " ... of a special grade partially designed for long circuits. It was imported from Germany where it was manufactured under the specifications adopted by the company as a standard."[7] The first telegraph line from Montreal to Vancouver was all iron wire. Strategically located repeater stations were required to boost the signal. Major repeater stations were built at Fort William, Ontario; Swift Current, Saskatchewan, and Donald, British Columbia.

Experiments with copper wire had been done in England about 1860. It was a superior electrical conductor sending a signal further before a relay would need to boost it. But copper was difficult and expensive to produce until 1895. At this time, copper had become less expensive to manufacture, could be rolled onto large spools, and was strong enough to withstand severe weather conditions. The first used diameter was 0.114 inches weighing 100 pounds per mile. A second preference was 0.104 inches in diameter weighing 172 pounds per mile. When tested in Canada under various weather conditions, the CPR found that hard drawn #6 copper wire (though mistakenly called #5) at 300 pounds per mile, or #9 copper wire at 210 pounds per mile would enhance high grade transmission more effectively. About 1898, Canadian Pacific began replacing the galvanized iron wire on their mainline circuit between Montreal and Vancouver with #6 hard drawn copper. This gauge withstood the severe weather conditions of the Rockies and Selkirk Mountains, and throughout Northern Ontario without causing telegraphic interruptions. By 31 December 1928, the CPR had converted 61.2 percent of their 160,287 miles of iron wire to copper; the remaining 38.8 percent remained iron. Canadian Pacific continued to replace iron with copper wire in an ad hoc manner; by 1937 copper wire was used on 65 percent of their circuits. Branchlines and more remote regions were changed to copper on an "as needed" basis. Many iron wire circuits were not replaced.

Other construction considerations included: length of the line, kind and size of wire, number of stations to be connected, style of equipment to be used, transmitter equipment, receiving equipment, induction coils and circuit, and amount of current needed.

FIGURE ONE: Weight Per mile Comparison of Galvanized Iron Wire and Hard Drawn Copper Wire

Galvanized Iron Wire	Pounds Per Mile	Hard Drawn Copper Wire	Pounds Per Mile
#14	97	#20	16
#12	163	#16	41
#10	264	#12	104
#9	323	#9	210
		#6	300

(Occasionally, Hard Drawn Copper Wire may appear as Hand Drawn Copper Wire, however, it is the same wire.)
- compiled from various sources -

It is noteworthy that copper wire rarely caused operational problems. Copper wire lasted at least fifty years before replacement was needed. However, indirect influences have hindered railway and telegraphic services. The theft of #9 gauge copper wire has been a problem, especially when the economy is bad. For example during the Depression miles of copper wire was stolen from the railway telegraph poles because it obtained a high price when sold. (Even as late as 1991 on Canadian Pacific's Galt Subdivision between Cambridge and Guelph Junction, thieves removed about twenty pole lengths of bare copper wire. It was estimated the thieves might net between $8,000 and $10,000.)

Insulators are essential to a proper wire circuit. Their primary function was nonconductive, therefore preventing electrical current from escaping. The earliest insulators were made of cheap bottle glass. An individual iron, copper or bronze wire is secured to the insulator. The insulator has interior threads that mate with the threads of the wooden pin or bracket. Both are secured to a horizontal crossarm either by nail, metal screw or wire. Glass was the best insulator (glass bottle necks, glass door knobs, manufactured glass insulators), however, rubber, phenolic, and Pyrex were also used. Porcelain insulators were utilized, although, in wet weather they tended to allow heavy electrical escapes; if located in dry areas such as under eaves, they worked properly. Another form of insulator was the rubber-hook style. A rubber-hook insulator could be found around wooden stations or buildings where wood work was extensive. It had a threaded shank that screwed firmly into the inside of the insulator. It attached easily to the side of a structure with two holes no larger than one-half inch each. Interestingly, not until the 1870s did Canadian insulator manufacturers begin using threading inside the insulator so they could be screwed onto a peg then attached to a pole. Until this time insulators had a threadless smooth inside. This style of insulator was affixed to a pole by glue or wire.[8]

A coloured insulator indicates impurities in the manufacture process of the glass. They range in colour from near-green to near-blue. Around 1915 selenium and cerium were added as a clarifier to the manufacturing process to remove the green tint caused by the iron oxide impurity in the glass. These insulators changed to a straw or light yellow colour when exposed to the sun. Manganese was experimented with from about 1880 until World War One when it was discontinued. Manganese was used to clarify the colour. When exposed to the sunlight insulators with manganese turned a purple or amethyst colour depending on the amount of manganese in the batch. On a pole in the 1850s to 1920s, one might see an array of colours in aqua, purple, clear, porcelain white, straw or yellow and even black rubber ones. The colour did not effect their insulating capability.

Canadian National Railways poster reward from June 1934. Insulators more recently have become collector items. Before, they were vandalised by stone throwers and target practice.
(6218 Museum, Fort Erie, Ontario, GTR Ridgeway Station / R.G. Burnet)

Photographs of telegraph poles, wires and circuits are rare. One has to rely on pictures such as this one from ca1900, CPR Caledon, Ontario station, to locate specific details. In particular, notice the poorly defined shape of the nearest pole, often used in rural areas . The wires came into the station around the rectangular window, connected to the telegraph key, sounder and possibly relay, pass through the freight area and exit out the back. Note the uneven distance between crossarms along the line.
(Courtesy Canadian Pacific Limited, Image No. A-12248, Photographer J.W. Heckman)

A fine example of a mainline telegraph pole with crossarms filled with insulators and wires. Note the 'roofed' top of the pole to shed rain and snow. The many styles of insulators in common use is prominent. On the pole, wires would be for the dispatcher's circuit, commercial operations (news papers, commercial telegraph, stock tickers etc.) and railway signalling circuits.
(CNR Public Affairs, #23112)

An interesting problem with insulators helped to improve their manufacture. At first it could not be determined why insulators were found smashed on crossarms. Some proported theories included hunters using the insulators for target practice, and kids smashing them with stones. While vandalism did contribute to smashed insulators, it was relatively small. However, it did become known that most broken insulators was due to expanding or contracting temperatures of trapped air bubbles inside the glass and this caused them to explode. Later, this problem was virtually eliminated with better manufacturing standards by Dominion and other companies.

Insulators, like copper wire, were targets of vandalism. Today insulators are collector items because of their colour and, telegraph company or railway names manufactured on them. They can sell from $2.00 to $10,000.00.

XI.
TELEGRAPH POLES

The earliest poles were crude and of poor quality. Trees were first used to secure the wires but it was only temporary. "Geore [sic] B. Prescott [was] the leading telegraph engineer of the time. The lines were usually built in haste, the posts generally set while filled with sap, often without taking off the bark. Consequently, in a few years, they rotted off at the surface of the earth, and then had to be replaced by others in the same manner."[9] These poles might last fifteen to twenty years. Later, dried and cured poles treated with creosote, painted or heavily coated in fat or oil were used. Unbarked cherry trees were tried along with: red cedar, white cedar, walnut, poplar, white oak, black oak, post oak, chestnut, honey locust, cotton wood, sycamore, birch, blue pine and scotch pine. Each tree type was tested for its strength against various weather conditions.

Poles were typically placed about thirteen feet away the nearest track with the crossarms no more than eight feet from the nearest rail. The poles parallelled the tracks but had to follow the contour of the land, and maintain uniformity as much as possible. Early pole height ranged from eight to thirty-eight feet. Later, poles ranged in heights of 16, 20, 25 and so on to 60 feet. They were classified according to their diameter and the crossarm weight they would need to support the wires. Pole classes ranged from one to ten. A class one pole was the strongest, had the widest diameter and was tapered to the top. A class ten pole was the smallest height and diameter and reserved for station or rural branchlines areas and generally not straight.

Poles were located according to land contours. A relatively even and consistent height aided electrical impulse, used less wire if it was kept at a constant height and aesthetically more appealing. There were at least six conditions to consider:

> One: hilly terrain - poles are placed on the top of hills, rises, span valleys and ravines, over creeks and rivers;
>
> Two: direction change - poles had to be anchored when the direction change was radical, such as 90 degrees, or when following a curve, poles needed anchors and down-guys to compensate for the inward pull of the wire;
>
> Three: poles may not necessarily follow the tracks due to tunnels, rivers or other serious obstacles, since they can be maintained within range of the tracks (this was vital in case of an emergency, a portable key could be attached to the wire);
>
> Four: crossarms ought to alternate from front to back on the pole to maintain a straight pole, and,
>
> Five: vertical clearance of eighteen feet over roads, highways or where other vehicles might cross under the wires and 27 feet or more over all railway tracks.
>
> Six: poles must, whenever possible, pass behind the station or other railway structures or on the opposite side of the tracks.

Pole tops had to be considered. Tops were either gained or roofed. This was done to provide drainage for snow, rain, and to prevent accelerated rotting of the pole from the top.

Early poles had a threaded wooden pin nailed to the top of the pole with the insulator. As more wires were attached to the poles, horizontal crossarms developed to distribute the weight of the wires more evenly and, making it easier to service wires and insulators. Notches were cut into the sides of the pole to provide support and stability to crossarms. Notches were generally cut about one inch deep then anchored with a nut, washers and bolt. The crossarms had to carry the weight of the wires. Oak, the preferred choice of wood, withstood the most severe weather conditions and wire weight; for instance, 210 pounds per mile copper wire required poles set from thirty-five to fifty poles per mile. The mean temperatures, maximum winds, snow, rainfall, and the gauge of the wire had to be determined to avoid damage to the circuit. Major mainline railways might use five to seven crossarms; heavy branches might have two or three, and lighter branches maybe one crossarm with six to ten pins.

On the crossarm, the first circuit was always the dispatcher's; a message circuit followed by a block circuit for station to station use, and possibly a local business circuit. Many other circuits were used, such as stock tickers, dedicated international press services, emergency circuits and telephone circuits when they became more prominent.

Placement of pole spans depended on the land contour and weight of the wire. Poles might be placed from fifty to one-hundred-and-fifty feet apart making it 105 to 35 poles per mile respectively. The average distance between poles on most Canadian lines was one hundred and fifty feet or 35 poles per mile; but varied considerably because of terrain and weather patterns.

Flagmen, according to the 1911 Grand Trunk Railway Rule 99 for instance, based their front and rear protective flagging distance entirely upon the number of poles and the standardized distance between them: 500 yards or 10 telegraph poles; 1,200 yards or 24 telegraph poles, and 1,800 yards or 36 telegraph poles (a distance of 150 feet in each case between poles). The employee timetable had the final say based on territorial terrain, and differed greatly between British Columbia, the Prairie and Maritime subdivisions.

This photograph is quite unique. Of immediate interest are the insulators and how they were secured to the pole by two large nails. Note the wire spacing and loop around the corner (which would never be allowed today), also the crude anchoring poles. The lineman has two spikes secured to his shoes with buckles on his calves to hold them in place. He has an axe in his hand, likely for shaping the top of the pole and is not using any safety equipment - see Appendix Three dealing specifically with these points. The photograph was taken in the Yukon about 1899 primarily for communication to and from the gold fields and connected with circuits at Quesnel, British Columbia.
(National Archives of Canada, PA-95736)

The lineman is adjusting and splicing the telegraph-telephone circuit. The insulators are on one style of rack used to prevent induction (interference) of telegraph and telephone circuits. At his feet are hydroelectric insulators. The top of the pole is gained. It is interesting to compare the previous photograph with this one. Note that now safety boots and belts are used, insulated gloves, wire insulators and a tool belt, none of which are present in the previous photograph.
(CNR Public Affairs, #X24678)

A wonderful depth-of-view photograph of the CNR mainline between Montreal and Toronto, possibly in the Cobourg area. The double crossarms are used at junctions, crossing over tracks, and for added support to the pole. On the right side of the photograph are the telegraph poles of the Canadian Pacific Railway.
(CNR Public Affairs, #41598)

Three men digging a telegraph post hole between Atlin and Quesnel, British Columbia ca1890. Note the simple tools for what must have been back-breaking work in tough terrain. This circuit, when completed, connected Quesnel to the Yukon by 1899 and used primarily during the gold rush years.
(CNR Public Affairs, #75119)

Pole and wire construction required tough and dedicated men:

> Largely French Canadian, possessing the pluck and hardihood of their class, these gangs of men made their way through swamp and forest, through town and hamlet, thinly clad, simply fed, performing their arduous work, in reliance upon orders from headquarters, with a rough strength, a sturdy loyalty, and a celerity that was almost awe-inspiring to the newly appointed local agent or raw operator, who for the most part felt his suddenly-acquired importance eclipsed by that of these Rembrandtesque sailors-of-the-forest.[10]

Pole construction in the far north was difficult. The CNR used tripod telegraph poles where perma frost existed, as did the Temiskaming and Northern Ontario (later Ontario Northland) Railway.

> In the perma frost, muskeg and tundra country, three telegraph poles are tied together to form a tripod, which cannot tip over for a single pole placed in the ground would not remain upright because of frost. This may be the only railway in North America along which telegraph poles are set as tripods.[11]

This method prevented the poles from toppling over and heaving in spring, and is still found along the Hudson Bay and ONR lines today.

XII.
POLE AND WIRE PROBLEMS

Westward construction involved other problems for telegraph pole construction. Prairie construction was characteristically straight forward. However, in the mountains and through northern Ontario, construction was more difficult. For instance, the blasting of the right-of-way often removed several poles and wire already set in place. This not only delayed telegraph services but obstructed train movements and ordering of necessary materials and supplies. Repair work " ... for a repairer (the location of a break having been previously ascertained by galvanometer) to reach the spot in an hour or two by jumping on a track velocipede [was not possible in the mountains]. Long and weary riding, climbing, wading was necessary, and the poor repairer was many a time 'desolate' when he reached the break."[12]

There were several problems interfering with telegraphic service. One of the most common problems was the weather. In the winter of 1942-3, one of Canada's worst snowfalls occurred, crippling telegraph and telephone service along the Montreal-Toronto tracks. An area of 9,000 square miles was affected, with 10,000 miles of telegraph wires and 1,500 poles totally or partially destroyed. Poles snapped because of the snow and ice accumulation on the wires and crossarms. Sub-zero temperatures also contributed to poles fracturing. Cornwall, the hardest hit, became an isolated community without communication of any kind for almost two weeks.

In severe weather, telegraph poles could sustain dramatic damage, disrupting communications for the railway. This example comes from the Cornwall area in 1942.
(CNR Public Affairs #50755-13)

Another freak weather occurrence in February 1942 caused a mysterious interruption to telegraphic service. While operating out of "C" office in Toronto, one wire after the next went suddenly dead. Between fifty and sixty relays stopped and by 11:30 p.m., no wires operated. After a telephone call to the wire chief, it was discovered that the aurora-borealis had caused the entire system to go dead. About 2 a.m., the wires started returning to life and working normally.[13]

Electrical problems abounded. It could be as simple as a maladjusted relay or a spring in a sounder armature coming loose or popping out. Wire corrosion, batteries dying and wires coming loose were quite typical. Another problem was rubber insulated telegraph cable. This material tended to create static charges that could send a random and sudden high static voltage through a relay station, with the potential of blowing up the building or causing a major fire.

Other problems included: brush fires, floods, sink holes and muskeg swallowing whole telegraph sections, moose and caribou catching wires in their antlers and dragging the wires through the brush. Bears were known to climb the poles in British Columbia and Northern Ontario and break the wires. Birds built nests around the insulators and pecked at the wire until it snapped. Thieves stealing copper wire, and insulator collectors were also quite common telegraphic difficulties.

CHAPTER FIVE

TELEGRAPH CIRCUITS

I.
SINGLE OR CLOSED CIRCUIT

Electrical telegraph circuits between 1840 and as late as 1870 operated one way. The single or closed circuit involved one key and message being transmitted at a time. All relays on the line would be activated simultaneously due to the alternating cessation and flow of the electric current. The basic circuit was simply operated: the key makes and breaks the electrical battery source that causes pulses of current to pass along a wire, and may require boosting with a relay to actuate the sounder at the receiving end. Technology at this time did not exist to individually signal one station; all stations were signalled and all listened in on the conversations, not unlike a Citizen Band (CB) radio today.

II.
SIMPLEX - DUPLEX - QUADRUPLEX CIRCUITS

Simplex was developed in the 1860s to improve telegraphic service. Simplex was a circuit that operated from both ends of a terminal with intermediate stations. The circuit, devised with early work on the telephone, had a ground-return telegraph path derived from a working metallic telephone wire and all without interference of signals. However, it did not work simultaneously.

Simplex had two useful conditions: one, during the idle condition, no current flowed along the circuit; two, the closed-loop system where current flows along the wire. A sending-receiving switch was operated before sending could occur. Also involved was a single current where there was an on/off signal. The double current allowed for the reversal of the signal making this the better of the two choices. It required a polarized receiver sensitive to current direction rather than its strength. As the number of messages increased, a greater need for a flexible simultaneous transmission in both directions was demanded.

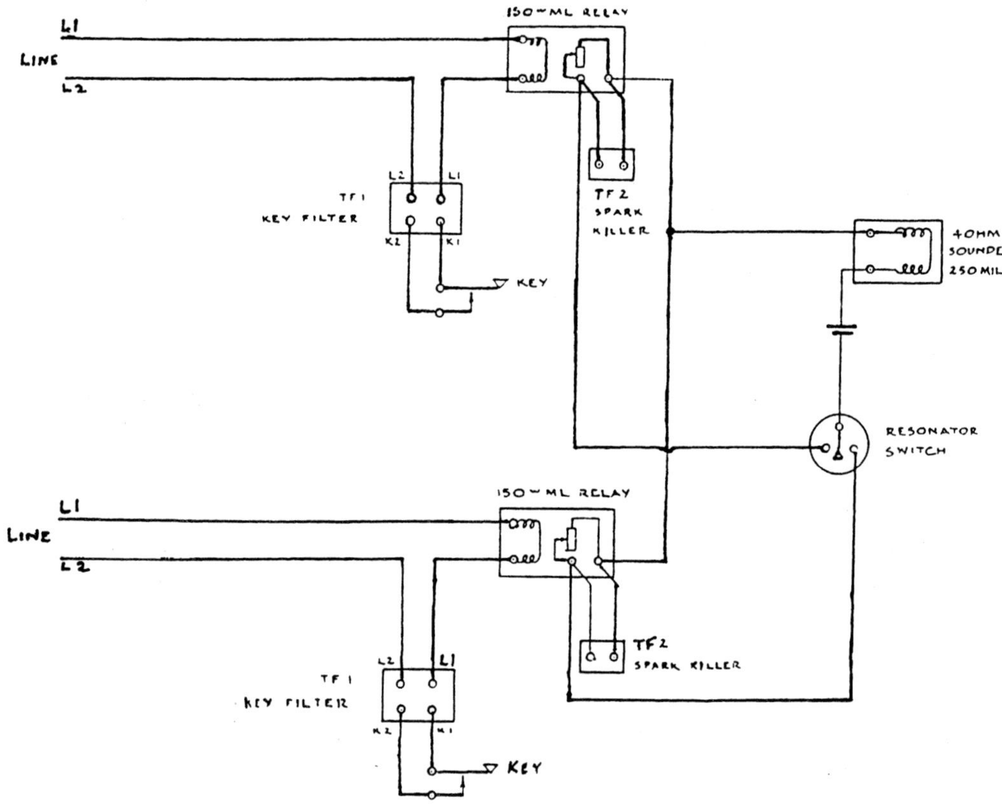

MORSE TELEGRAPH SETS

TWO LINE

LS 2

Schematic diagrams from the CNR Communications Department for specific telegraph set up. (CNR Communication Department, courtesy G.W. Horner)

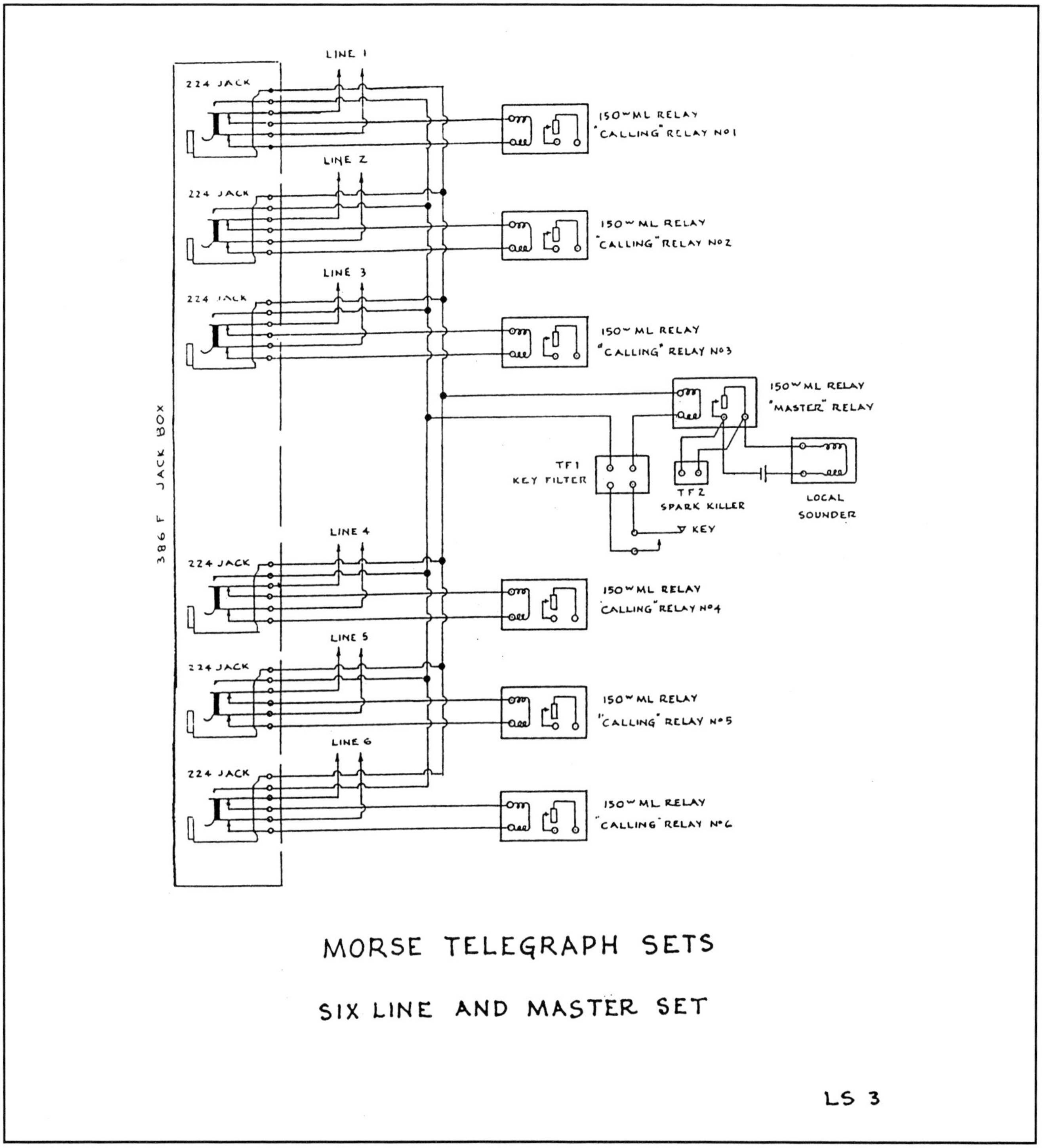

Schematic diagrams from the CNR Communications Department for specific telegraph set up. (CNR Communication Department, courtesy G.W. Horner)

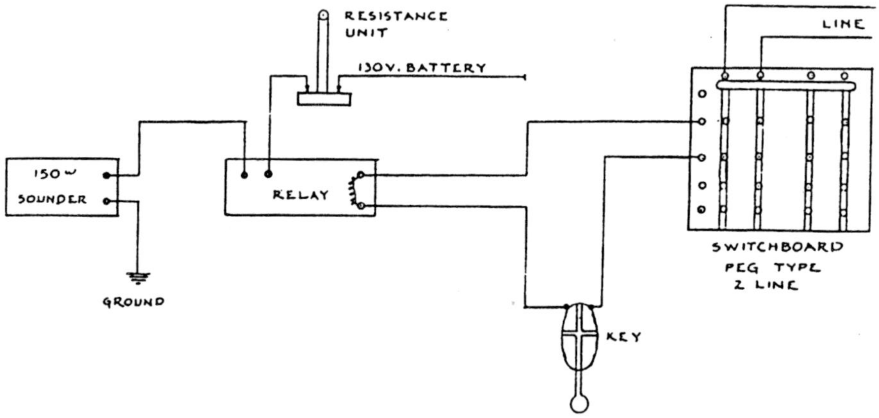

BLOCK DIAGRAM

FOR MORSE OPERATION

130V. LOCAL BATTERY

LS 4

Schematic diagrams from the CNR Communications Department for specific telegraph set up.
(CNR Communication Department, courtesy G.W. Horner)

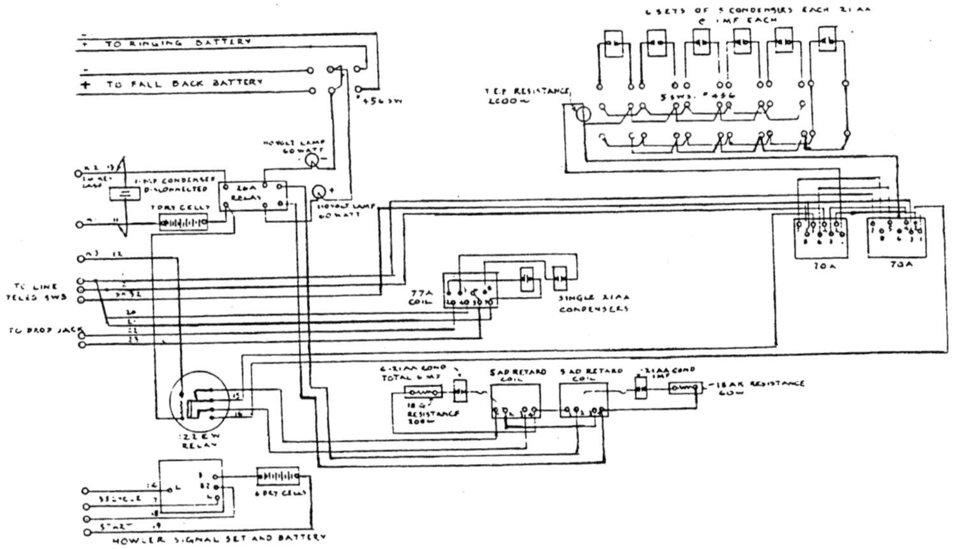

WIRING DIAGRAM

OF A DISPATCHER'S OFFICE

LS 8A

Schematic diagrams from the CNR Communications Department for specific telegraph set up.
(CNR Communication Department, courtesy G.W. Horner)

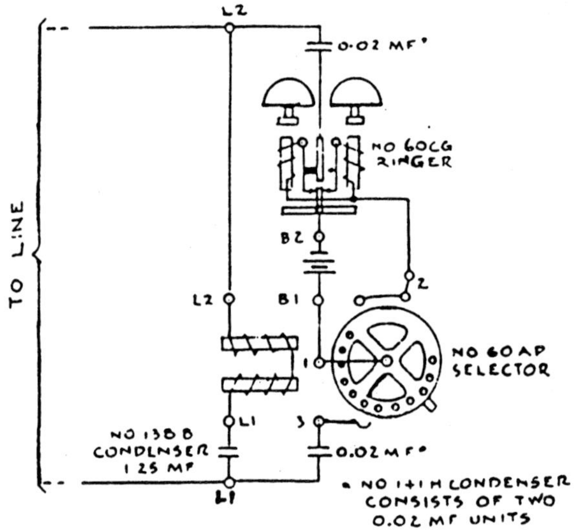

NO. 160C SELECTOR SET

SCHEMATIC

LS 9

Schematic diagrams from the CNR Communications Department for specific telegraph set up. (CNR Communication Department, courtesy G.W. Horner)

Duplex or "Differential Duplex Circuit" evolved. Duplex was theorized as early as 1853 by Wilhelm Gintl. The duplex circuit permitted for the first time two simultaneous transmissions in opposite directions between two terminals. The principle involved an interconnection of the transmitting and receiving units.

Later, a balanced network was required so the receiver would not be affected by the sending electrical current. A variation on duplex was 'Bridge Duplex' which was similar to differential and did not add any significant advantages. About this time, selective ringing equipment began to appear. This permitted a dispatcher to signal and send a message to an individual or series of selected stations without disturbing others. However, it was not completely used by the telegraph until the telephone perfected the technology.

Thomas Edison invented quadruplex. This system was a combination of duplex and a modified single-current duplex. Quadruplex could provide four messages simultaneously in both directions. This meant two receiving and two sending devices were used. One was a polarized unit responding to the direction of the current; the other was nonpolarized and responding to change in current magnitude, irrespective of the electrical impulse direction. It was a terrific advancement but only an intermediate step to more efficient devices.

Another circuit type explored in the 1890s was the derived or superimposition circuit. This evolved for combined telegraph and telephone use on the same wire. Telegraph current is divided so half the current passes over each wire. No interference is felt by the telephone as the current is equally divided so the two halves neutralize each other's magnetic effect.

Multiplexing or Time Division produced the most effective results. It had evolved from Baudot's multiplexing sixty years earlier. This involved the division of the circuit among several operators. Each operator would have their own channel on that single wire. Special devices were created to maintain accurate synchronization and a correct electrical phase between sending and receiving. It worked by allotting cyclical connections to the transmission line for a given time. It was not extensively used until about 1930.

III.
METALLIC AND GROUNDED CIRCUITS

A 'metallic circuit' consisted of two wires. One wire was for the main, the other, return. With only one main wire, several instruments could be connected to it and not present operational problems. Similarly, several instruments could be connected to one return wire. However, care was necessary to properly insulate each wire from the other. This type of circuit was best used on short lines. Metallic circuits were generally not connected to earth return.

A 'grounded circuit' differed to a metallic circuit by having a single main line with earth used as the return. Since both ends of the grounded wire were attached to the earth, the current flowed readily and efficiently. The mileage for such a circuit was immaterial. This circuit style was preferred by many companies, saving expenditures on construction and wire.

IV.
CARRIER CURRENT

Of all previous circuit inventions, carrier current far exceeded expectations for railway communication. Carrier Current was developed around 1920. It could support both telegraph and telephone communication. The principle involved alternating rather than direct current. Carrier Current was achieved by varying the amplitude of the carrier wave along the wire by switching it on and off, or altering the phase of the carrier. Another wave could be initiated by modulating the amplitude to a slightly different frequency or phase. Tones of different frequency or pitch were used to carry the telegraph signal. This occurred simultaneously over a pair of copper wires.

Carrier current allowed and supported several different frequencies on the same wire using many distinctive channels. Spacing between signals depended upon the required speed of transmission. At first, the most common spacing was 120 cycles per second resulting in 12, 18 or 24 channels, allowing conversations in that number to occur over a four-wire circuit. Each channel was controlled by an independent transmitter on the same wire. Electrical filters removed distortion and kept the signals pure until they reached their receiving unit. The signal was then filtered and demodulated by an amplifier detector. The frequency of the carrier was technologically restricted to a range of 300 to 11,000 cycles per second. Telegraph carrier channels were divided between 300 and 2,600 cycles per second (cps) by using bandpass filters, while the telephone carrier used between 3,700 and 11,000 cps. Telegraph or telephone signals could also operate between 2,600 and 3,700 cps.

Canadian National Railways inaugurated carrier current on 26 April 1927 between Montreal and Toronto. Their initial service carried ten Morse code cyphers and two telephone messages over the same wire simultaneously. It "... was said to have been the first occasion on which these forms of communication were combined [that is Morse and telephone circuits]."[1] Canadian National moved quickly to install carrier current (Figure One); Canadian Pacific followed (Figure Two).

FIGURE ONE:
Carrier Current Installation - Canadian National Railways

Completion Date:	Location:	Channels:
April 26, 1927	Toronto to Montreal	10
November 1927	Toronto to Winnipeg	6
December 1927	Winnipeg to Saskatoon	6
January 1928	Saskatoon to Edmonton	8
March 14, 1929	Edmonton to Vancouver	8
August 1929	Toronto to Montreal	upgrade to 20
September 1929	Winnipeg to Saskatoon	upgrade to 10

- complied from several sources -

FIGURE TWO:
Carrier Current Installation - Canadian Pacific Railway

Completion Date:	Location:	Channels:
October 2, 1928	Inauguration of Service Toronto to Montreal	10
March 25, 1929	Toronto to Winnipeg	10
September 30, 1929	Winnipeg to Vancouver	10
April 15, 1930	Montreal to Winnipeg	8
December 1, 1930	Montreal-Toronto-Windsor	10
October 1931	Toronto to Windsor	10
March 1, 1932	Montreal to St. John, N.B.	10
June 1, 1932	St. John to Halifax	10
December 1945	Montreal to Winnipeg	upgrade to 10

- complied from several sources -

Carrier current contributed to the telegraph's demise. Carrier current cut the number of telegraph facilities already in place. Because many more transmission channels were possible on one wire, thousands of miles of telegraph lines could be removed. Carrier also produced a tremendous increase in exceptionally faster telegraphic traffic. By 1937, Canadian Pacific had developed carrier current to handle thirty-six separate telegraph channels along two copper wires used primarily between

Toronto and Montreal. Some commercial companies were able to develop "... frequencies up to about 150,000 cycles on open wire lines and up to about one million cycles over specially constructed cable circuits."[2]

With carrier current, some distortion developed. Signals received were generally effected by either an increase or decrease in the signal strength. If this distortion was too great, code confusion resulted. Relays however could be adjusted quickly enough to allow for variances. Occasionally, the relay became maladjusted making the line useless for transmission until the proper corrections were made.

CHAPTER SIX

TELEGRAPH PEDAGOGY

I. DEVELOPMENT OF MORSE CODE

Encrypted signals can be traced back many centuries. Recorded by the Greek historian Polybius in 300 BC, the twenty-four letters of the Greek alphabet were encoded for signal messages. A large wooden box containing five horizontal and five vertical wooden slats were manipulated into various codes by turning the row-column slats. The Romans used a five-by-five-grid pattern that combined two patterns to indicate one letter. A precursor to Chappe's code was suggested in 1551 by Gerolamo Cardano, an Italian mathematician. His coding invention had five fires burning on five high towers; by blanking out various flames numerous codes could be transmitted. Francois Bacon, in 1605, developed the first comprehensive but awkward dot-dash code while studying cryptography. This method consisted of one unique combination of a dot or dash for each letter. For the next two hundred and thirty years, many different code styles evolved, but most were difficult to remember and apply effectively. Morse and Vail however would succeed with their final encrypted code version because of the electromagnetic telegraph.

Three types of code are associated with Morse and Vail. In 1835 the first code developed by Morse was the Double Translation Code. This method involved the conversion of words into numbers before transmission. The coded numbers were then transmitted to the receiving operator who then read the punched recording tape produced by the register back into words.

Morse published a number-code book to assist encryption with the Double Translation Code. Words were given a numerical reference not unlike a mileage sign. Numbers one to five were each given their respective numerical dot followed by a time-delay space equal to three silent dot units; six through zero used the same dot sequence as one to five adding a dash at the end followed by six silent dot units. The time-delay between and after numerals was the key to a successful transmission of the code. Though the principle was rudimentary, a simple sentence could become a numerical nightmare with great delays in referencing and decoding the sequences from the Morse Dictionary. Notwithstanding, this method was discarded because many words were not included in the Morse Dictionary and could not be decoded or encoded. Other reasons encompass extensive reception problems with the circuit and mechanical faults of the receiving paper tape register.

FIGURE ONE - FIRST MORSE CODE DEVELOPED - ca1835[1]

NUMBER	CODE	NUMBER	CODE
1	.	6	.-
2	..	7	..-
3	...	8	...-
4	....	9	....-
5	.....	0	.....-

Example: the numeric code for 'alarm' was:

...[3tds*]..[3tds]..[6tds]

*** tds = time-delay space equal to three or six silent dots**

The second method was devised in late 1837. Alfred Vail has been credited with this method. It is believed Vail visited a printer's type setting shop to learn the frequency of the letters used in preparing published newspapers and pamphlets. Vail observed and concluded that the letter 'E' was the most common letter in the English language, so he assigned a single dot (dit); incorrectly, he believed 'L' was second, and it received a single dash (dah) while the letter 'I', third most frequent, received a single dot and dash, forming the dit-dah sequence; 'Q', the most infrequently used letter, was combined into dit-dit-dah-dit. In this manner, every letter and numeral had a dot-dash code. A significant change with this code version was the time-delay space. The time-delay now occured after letters, words and sentences: between letters there were three time units or the value of three silent dits and six time units between words or the end of a transmission. (Sources conflict on the delay time period however most state the time of six silent dots was practiced.)

FIGURE TWO - SECOND CODE DEVELOPED - ca1837[2]

LETTER	CODE	LETTER	CODE	LETTER	CODE
A	...	J	.. .	S	.-.
B	.. ..	K	-.-	T	--.
C	. ..	L	-	U	.--
D	... .	M	-..	V*	-
E	.	N	-.	W	..-
F	. ...	O	..	X*	-
G	.. .	P	.....	Y	.-
H	....	Q	..-.	Z	.-.
I	.-	R	. .		

***Some sources use a solid double dash for V and X**

Example: 'alarm set'

A	L	A	R	M
... [3tds]	**- [3tds]**	**... [3tds]**	**. .** [3tds]**	**-.. [6tds]**

S	E	T
.-. [3tds]	**. [3tds]**	**--. [6tds]**

**** space between dots as in B, C, D, F, G, J, R equal to one silent dot**

A serious flaw became evident with this system. By examining Figure Two, G-J, I-Y, V-X, and S-Z had the same dot-dash sequence. The receiver could only determine by context which letter ought to be used after the transmission had terminated.

It was not until two weeks before the famous May 1844 transmission that Morse wrote a letter to Vail. The letter expressed Morse's concern over electrical failures along the line and the faults in the code they were testing. Morse requested that Vail make a longer pause between each letter, lengthen the space duration between words and, developed a new pattern for those letters with the same dot-dash sequence. Morse also pointed out the frequency of letter errors with respect to 'T' and 'I', and this was resolved in the next code chart.

Figure Three, given out by the Canadian Pacific Communications Department to agents and operators, details the standardized Morse Code used by the railways.[3]

CANADIAN PACIFIC RAILWAY COMPANY

COMMUNICATIONS DEPARTMENT

MORSE CODE

ALPHABET / NUMERALS

Letters	Morse	Letters	Morse	Figures	Morse
A	•-	N	-•	1	•--•
B	-•••	O	• •	2	••-••
C	•• •	P	•••••	3	•••-•
D	-••	Q	••-•	4	••••-
E	•	R	• ••	5	— — —
F	•-•	S	•••	6	••••••
G	--•	T	-	7	--••
H	••••	U	••-	8	-••••
I	••	V	•••-	9	-••-
J	-•-•	W	•--	0	—
K	-•-	X	•-••		
L	—	Y	•• ••		
M	- -	Z	••• •		

.	Period		••--••	£	Pounds	LX	— •-••
:	Colon	KO	—•— • •	/	Shillings	UT	••- -
: -	Colon Dash	KX	—•— •—••	d	Pence	D	—••
:"	Colon Quotation	KQ	—•— ••—•		Capital Letter	CX	•• • •—••
;	Semicolon	SI	••• ••		Small Letter	I5	•• ---
,	Comma		•—•—		Decimal Point	DOT	—•• • •—
?	Interrogation		—••—•	%	Per cent	OSO	• • ••• • •
!	Exclamation		— — — •	¶	Paragraph		— — — —
–	Fraction Line	E	•		Not Code	E5	• — — —
/	" "	UT	••— —		Underline	UX	••— •—••
—	Dash	DX	—•• •—••		Underline (End)	UJ	••— —•—•
=	Double Dash	BK	—••• —•—	(	Parenthesis (Start)	PN	••••• —•
-	Hyphen	HX	•••• •—••	)	Parenthesis (End)	PY	••••• •• ••
'	Apostrophe	QX	••—• •—••	[ ]	Brackets	BX	—••• •—••
$	Dollars	SX	••• •—••	"	Quotation (Start)	QN	••—• —•
¢	Cents	C	•• •	"	Quotation (End)	QJ	••—• —•—•

Repeat when sending	DN	—•• —•	Break or stop sending	BK	—••• —•—
I understand	13	•——• •••—•	Finish (No more)	30	•••—• ——
Message to all offices	23	••—•• •••—•	Regards	73	——•• •••—•
I am busy	25	••—•• — — —			

Figure Three was issued by the Canadian Pacific Railway Communication Department to agents and operators as a reference only. Compare this Figure with Figure One and Two: note that specific numbers and letters remained unaltered. Subtle changes were made however to avoid conflicting code. (W.J. "Jack" Corbett)

As a matter of instruction, the dash (dah) was standardized at three times the length of the dot (dit); the space between a dit-dah was equal to one silent dit in time. Spaces between a character were equal to the time of one dit; the space located between characters was equal to about three dits, and the time between words was seven dits in length. However, all this is relative as operators with great skill could far exceed these instructional times.

A need for code standardization became a necessity around 1880. Railway cars from one road interchanged with another railway were occurring more frequently, as was international traffic between Canadian and American railways. To compound operating problems between Canada and the United States, "... some railways insisted on their own code for little better reason than to insure their singularity."[4] For each railway and commercial company that existed, there tended to be a unique and separate code. Secrecy in business was paramount and secret codes were essential to railway and business survival. For example, where one sequence of numbers, such as sending a '999' might mean "Stop train. Emergency"; the same sequence with another railway might suggest "Station being robbed. Send police." It was argued, money, safety and speed would be gained by standardized codes.

Standard American Morse code was therefore adopted for international railway use between Canada and the United States by 1889. The General Time Convention, later the Association of American Railroads, standardized the meaning and use of American or railway Morse code. The established and standardized Railway Code also included uniform train operating rules and the movement of trains through telegraphic train orders.

Rulebooks developed also. They would vary from railway to railway and one edition to the next but the general meanings were the same. Before 1889, rules and codes were subject to wide interpretation. However, after standardization a Superintendent of Examinations and Rules became the judge and interpreter when rules were violated or misinterpreted by an employee. If a judgement was levied against an employee, Educational Demerits were deducted according to the severity until the maximum points for dismissal was reached. Some errors, such as failure to stop a train at one's station meant instant dismissal. Secret company coding would remain, however, it took on a different encryption of letters and numbers but the cryptography would not compromise international, commercial traffic or the safety of employees, passengers and equipment.

II.
TEACHING CODE TO RAILWAY OPERATORS

Sound reading was discouraged in the early stages of Morse telegraphy. In 1846, Morse was influential in having the US Government pass a law forbidding the practice of sound reading. Morse obviously had anticipated sound reading but failed to realize its efficiency - or he did and wished only to protect the interests of his paper tape register. Thus, for many years, railway circuit managers and wire chiefs continued to insist "... the letters [be] decoded from the inked paper strips."[5] No such similar law is known to have existed in Canada. However, early company policy may have demanded it, otherwise, the purchase of paper tape registers was ineffectual.

The consensus for teaching and learning code became 'reading by sound' after 1856. Jimmy Leonard is credited with discovering this method. Jimmy realized the telegraph was his future. He began working in the fall of 1848 at the St. Clair Street Telegraph Office in Kentucky. By age fifteen, he had become known as a fast, efficient sending and receiving operator. James D. Reid, the Louisville area manager, sent for Jimmy one day to commend him on his skill and requested a demonstration. Jimmy, with his sending friend Joseph Fisher, successfully demonstrated receiving fifty-five words in one minute completely by sound. The manager was horrified and reprimanded the boys for breaking the law.

The feat of the two boys however was brought to Morse's attention. What Leonard and Fisher had achieved was translating the coded sounds of the paper tape register into words, dismissing the decoding phase of the tape. Sound reading was their secret in achieving fast and perfect transmissions. Later, reason won and the law was repealed. Now, all practice and applications were to be by sound only; one was encouraged to hear the dit-dah rather than decode register tape.

Morse developed a "fast-tracking" mnemonic teaching method primarily for Civil War soldiers. The code consisted of syllabic, rhythmical word phrases. This code could be memorized "... in about an hour ...[rather than the] ten days to recruit signallers to learn the Morse Code .... This method was invented by Morse himself for use in the U.S. Army."[6] Nevertheless, the mnemonic code would take as much as two years to fully master to become second nature.

FIGURE FOUR - MORSE'S METHOD OF LEARNING CODE CA 1860[7]

A Ag-ainst : di-dah
B Bar-ba-ri-an : da-di-di-dit
C Cont-in-ent-al : da-di-da-dit
D Dah-li-a : da-di-dit
E (short) : dit
F Fu-ri-ous-ly : di-di-da-dit
G Gal-lant-ly : da-da-dit
H Hu-mi-li-ty : di-di-di-dit
I I-vy : di-dit
J Ju-ris-dic-tion : di-da-da-dah
K Kan-ga-roo : da-di-dah
L Le-gis-la-tor : di-da-di-dit
M Moun-tain : da-dah
N Nob-le : da-dit
O Off-ens-ive : da-da-dah
P Pho-tog-raph-er : di-da-da-dit
Q Queen Ka-tha-rine : da-da-di-dah
R Re-bec-ca : di-da-dit
S Se-ver-al : di-di-dit
T Tea : dah
U Un-i-form : di-di-dah
V Ve-ry va-ried : di-di-di-dah
W Wa-ter-loo : di-da-dah
X Ex-hi-bi-tion : da-di-di-dah
Y Youth-ful & fair : da-di-da-dah
Z 2 long/2 short : da-da-di-dit

The above alphabetically arranged words correspond to the precise number of syllables in the dit-dah sequence.

Many other styles of teaching and learning Morse code evolved. One method of instruction was rote. For example, the letter 'A' would be repeated often by sound, without variation, until its pattern was 'sound memorized.' Similarly the remaining code would follow. Constant review of a previous letter was done to reinforce the sound patterns. It was a tedious method, dull and uninteresting, yet it had its merits and worked for some people.

Railway training centres and independent Business Schools taught in groups. Group teaching had a double effect: more students could be taught in a class situation, and the code was taught in group patterns. The students learned letter and numeral code patterns in fixed clusters. The 'dits' and their closely related patterns were taught first since dot patterns far exceeded frequency over single dash patterns. For example, dit groups (eg) E, I, S, H, P (and) 6 were followed by various combinations of dits (eg) C, O, P. The following sentence and similar ones like it, were used to instruct dot codes with this method: SEE HER IRISH EYES. The entire sentence is a series of dots without punctuation as the period has two dots-two dashes-two dots in sequence. Also with this teaching method, the dash and all remaining combinations of dits and dahs were taught: (eg) T, L, M, 5 (and) 0. Finally, combinations of code for more complex meanings in sound and use were taught.

A variation of group instruction existed. Groups of letters with related sounds and similar characteristic code patterns were taught. Some teachers believed that pattern learning was easier. For example, the letter 'U' was di-di-dah; the letter 'Q' was di-di-da-dit.

Another method of instruction was teaching the most frequently used letters first, and the least last. Further, infrequently used codes were not taught. Therefore, the letters used more often in messages should be the most familiar to the operator. It could not be explained what the operator would do if an unknown code was sent. One suggestion was to either ignore the code or write down its dit-dah then look it up. Both methods were totally unacceptable to Canada's railways.

Interest of many young boys 'hanging around' the station probably played the largest teacher in learning the code. The agent might be a father, uncle or friend, and they would teach the son, daughter or young visitor railway Morse. Indeed, most operators, and some of the best, said that their father or grandfather taught them the code. Other operators might have student-operators assigned by the Company. They might be learning the ropes of station life from sweeping out the stock cars, handling baggage, filling out the endless reports while learning the key. Interestingly, some teachers of code felt that because life was unpredictable, and messages and their random patterns arrived over the wire, this was the best way to teach Morse. It was believed that learning in no specific order was another 'one of the best ways.'

Future operators could also attend specialized technical schools. There were many telegraphy schools outside the railway for instruction. One could learn telegraphy at Cassan Systems located on Queen Street in Toronto (they also offered telegraphy through correspondence). The Thomas Institute in Quebec City, and the EBI (or Elie) Business College in Montreal provided telegraphic instruction to hundreds of Canadian telegraphers for railway, wireless and newspaper services.

With the advent of World War One, many skilled railway telegraphers went voluntarily to Europe between 1915 and 1918. The Canadian Overseas Railway Construction Corps (CORCC) had two Telegraph Crew units. Their primary objective was to maintain and reconstruct damaged telegraph and telephone communication lines in the trenches. It also served the light railway system established in the trenches for the movement of ammunition, wounded soldiers and supplies.[8]

During the Second World War, the Armed Forces taught soldiers railway Morse, International and Phillips (Press) Code. The Royal Canadian Air Force, the Canadian Army and especially the Royal Canadian Signal Corps, taught their recruit's Morse. It must be noted, that many Canadian railway operators who enlisted in the Armed Services during World War Two, helped to revitalize the destroyed telegraph and telephone circuits throughout Europe. Even after the War, the Veterans Affairs Trade School in Toronto continued to teach railway and wireless Morse to returning soldiers in preparation for new jobs.

About 1945, flash card instruction appears. On one side of the flash card was the Morse code symbol in dots and dashes, and on the reverse side, its meaning. However, memorizing the character's visual pattern was not instructive. Using a chart meant wasted time in coding-decoding patterns and looking up each letters meaning or code. Operators who relied on a visual code chart were considered by most railway operators to be unskilled, slow and dangerous. It must be emphasized Canada's railways taught strictly by sound.

One theme throughout all methods was never to anticipate code letters or messages as a receiver; it was a dangerous practice to fall into. To teach avoidance of anticipation, well-known passages were often used then altered in many different ways. The student would decode the message on paper or use a typewriter. For example, one poem used went:

> I kept six faithful serving men,
> They taught me all I knew,
> Their names were "where" and "what" and "when"
> And "why" and "how" and "who."

Telegraph teachers new that poems of this sort had been drilled in school. To avoid anticipation the phrases were altered:

> I kept six faithful serving men,
> They taught me all I knew,
> Their names were "why" and "how" and "when"
> And "where" and "what" and "who."

Of course this had limited use and many other poems or phrases would be needed. All punctuation (quotations, semicolons, period and so on) was also required for coding/decoding.

For those young people with practice sets, learning Morse was not much different from railway instruction. Railway and learner manuals had four basic suggestions to success. Individual practice should always be maintained to become completely familiar with the Morse. Next, one needed to alternate key work or sending with another student or friend who listened or received. This would serve to correct inaccurate sending signals. Third, when it was felt they could carry on a simple Morse conversation of sentences separate practice in different rooms was encouraged. Finally, if possible, practice in an actual telegraph office was highly recommended.

As railway and commercial telegraph offices expanded throughout Canada, more operators and agents were required. The teaching of telegraphy grew rapidly and schools could be found throughout Canada. Professionally, the Canadian railways provided the best and most up-to-date methods to teach their employees. In every Canadian capital city, Canadian National and Canadian Pacific taught telegraphy to their new employees going into the telegraph department. If the would-be telegrapher passed a standard send/receive examination successfully, he/she was given a seniority date and placed on a probation-like period between six months and one year to ensure their capability.

FIGURE FIVE:
SIGNIFICANT CANADIAN RAILWAY TELEGRAPHY SCHOOLS

NATIONAL:

Canadian National and Canadian Pacific Railway Telegraph Departments in Halifax, Montreal, Toronto, Winnipeg, Saskatoon, Regina, Vancouver

NEWFOUNDLAND:

Newfoundland Postal Telegraphs Company St. Johns Night School

QUEBEC:

Canadian Army Operator's School - Quebec City
EBI (or Elie) Business College - Montreal
Number One Wireless School - Montreal
Thomas Institute - Quebec City

ONTARIO:

Bell Telephone Company, Morse Room - London
Canadian Vocational Training School - Toronto
Cassan School - Toronto (also known as: Cassan Systems, Cassan Systems Correspondence Department, Cassan Telegraphy School)
St. Thomas Collegiate Institute - Army Cadets
School of Telegraphy - Port Arthur
Telegraph School - Sudbury
Veteran Affairs Trade School - Toronto

MANITOBA:

Returning Soldiers Rehabilitation Centre - Winnipeg
Skinner's Telegraph School - Winnipeg
Western School of Railroading - Winnipeg

SASKATCHEWAN:

University of Saskatchewan

ALBERTA:

Alberta College - Calgary

BRITISH COLUMBIA:

Sprott Shaw Business College - Vancouver

ARMED FORCES

Royal Canadian Air Cadets, Royal Canadian Air Force, Royal Canadian Army, Royal Canadian Army Cadets, Royal Canadian Navy, Royal Canadian Navy Cadets, Royal Canadian Signal Corps

- compiled from many sources -

III.
PHILLIPS or PRESS CODE

In 1875, Phillips Code was developed. Because the volume of public information was increasing rapidly throughout the world, and worldwide telegraphic connections began appearing, more information was being sent. This code was advantageous as it provided a faster and more logical way to send out high volumes of daily articles, news events and columns. Because it was common practice for an operator to be commissioned, the more one sent the more one earned.

Abbreviated codes developed during the summer of 1875. Walter Polk Phillips, a high speed telegraph operator with the Associated Press (AP) in the United States, was assisted by four other employees of the AP office - Joseph Christie, Wilfred Gove, H.A. Wells, and W.H.C. Hargrave. They created this Code while on vacation in Cape May. Their objectives included making a code book that would speed up transmissions; remove the drudgery of copying press releases into long hand and sometimes making as many as ten copies from the same transmission, and, provide a faster mnemonic aid for code retention by operators. The book developed into a complete list of abbreviations derived from popular English expressions. In 1876, Phillips' book and code was published and adopted on the Associated Press circuit after careful testing between New York, Philadelphia and Baltimore. A revised abbreviation code book was published in 1879. Phillips Code became a standard for newspaper, industry and some commercial wire services.

It is noteworthy that Phillips developed a plan to use paper tape recorders and typewriters. The transmitting operator had his code copied onto paper tape; the receiving operator would then transcribe the tape. It was found this did not speed up the large quantities of news. With improvements made to the typewriter, Phillips successfully allied the receiving operator with a typewriter to the sending Phillips Code operator in 1883. So successful was this approach that it replaced the paper tape recorders almost overnight. Newspaper operators soon learned to translate the code while banging out the message with a typewriter at the same instant.

Figure Six gives several examples of the code structure.

FIGURE SIX:

PHILLIPS or PRESS CODE SAMPLING

Abbreviation	Word	Abbreviation	Word
ABB	abbreviation	PB	probable
AGR	agree	PD	paid
BF	before	R	are
BLU	blue	SNT	sent
C	see	TBL	trouble
HAP	happy	U	you
IW	it was	VY	very
KM	communicate	Y	year
LV	leave	YD	yield
NR	near	ZA	sea
OMN	omission	ZD	said
OT	out	ZM	seem

- compiled from various sources -

Phillips used single letter code abbreviations. He used the letters of the alphabet including 'A' and 'I' which represented separate but complete words. In total, twenty-one single letter codes were used by telegraph operators. However, multiple letter combinations did begin to appear. They were used to avoid the ambiguity some abbreviated codes presented as they could represent different phrases and therefore different meanings. Phillips believed combinations exceeding four-letters would cause problems when expanding the code by the receiving operator. This was not so: between ten and sixteen letter combinations were transmitted. Another practice with some companies and operators involved combined codes without spaces between words; for example, SNDPLCTBLNSTN or 'Send Police trouble in station.' A deviation also evolved where specific codes were assembled for an entire phrase: for example, PBOMN was used to transmit 'Probable omission in message please wait.' This new method was far superior, with more work accomplished. Phillips Code in Canada was used on commercial telegraph services only.

Phillips Code began to disappear when teletypes became more prominent in the 1940s. However, a variation of this code style exists today by many people using specific Internet Relay Chat channels from their home computer to rely information and "talk" around the world.

CHAPTER SEVEN

TELEGRAMS, SERVICES, FORMS, and EMPLOYEES

I.
TELEGRAMS

Sources argue the first telegram originated as early as 1084 BC. It is further recorded that beacon fires, located on high vantage point hills, relayed a coded message to Queen Clytemnestra's palace in Argos, Greece concerning the fall of Troy. Fire and smoke signals were used consistently for centuries by every nation on every continent to transmit messages. Other sources herald Napoleon using the Chappe system as creating the first telegram message while more recently, Morse's first official telegram, What hath God Wrought! is considered the true telegram. Whatever one's position the term 'telegram' is clouded by how we have used it in our recent past and early references called this a message, despatch, news, bulletin, or note. Technically, the word telegram is from the Greek: tele- meaning 'far' and -graphein (-gram) meaning 'thing written.' The definitive term 'telegram' is credited to the French inventor, Claude Chappe.

The telegram of the Morse creation and technology has served many purposes. Initially, the Morse coded telegram was used for commercial enterprises, businesses and newspapers. Other telegram uses included the purchasing, ordering and selling of goods, stocks and bonds. As the telegraph grew in significance, it became associated with railway signalling, train location, railway business, and correspondence. Then, it broadened out to include a social consciousness, conveying tragedies, declarations of war, reporting world events, and for social occasions such as invitations, greetings, announcements, births, deaths. (Indeed, today's World-Wide Internet can be traced to the world connections made by telegraph beginning over one hundred and fifty years ago.)

The railway took pride in their telegram service. Because the station agent was the frontline company employee, courtesy and knowledge were demanded. In 1911 the Grand Trunk Rulebook specifically stated that railway employees " ... will be held responsible ... for the courteous behaviour and proper deportment of all employees ...."[1] People came to the office or station and filled out their message over the counter. The agent had to know his tariffs, the time when the telegram would be sent, received and delivered. When the telephone became prominent, phoning in a telegram became popular. Canadian National demanded: "Success in business depends on how we treat our customers. It is as simple as that. So much business is done today over the telephone that a pleasing Telephone Personality is just as important as a Face-to-Face Personality. Anyone can acquire it. Cultivate the 'friendly' voice."[2] Also, one was to answer the telephone promptly, and "The first response should be 'Canadian National' ... listen patiently without interruption; [when the conversation is over] ... say a courteously said 'Goodbye' or 'Thank You."[3]

Dedicated telegram offices were located in most every town and city. Telegram offices were often combined with ticket offices in larger cities. These offices were generally simple yet ornate. Mahogany, oak and cherry wood the usual material inside. The front windows displayed the services offered to the public. Letters were generally gold painted with black trim in large upper case block letters. Sometimes the company logo might surround the word telegram.

Telegram competition in cities was often intense and telegraph companies would try to 'out do' the other. Locations were vitally significant, and many city telegram offices occupied a busy and prominent street corner. In cities like Toronto, Montreal and Winnipeg, the competition might be right across the street or on the same city block, even right next door. In rural areas, the railway station served as the centre for sending commercial telegrams while also handling railway telegrams. If a town did not have a railway station, a commercial telegraph office was available.

Interestingly, telegram-ticket offices where located around the world in major cities like Hong Kong, Tokyo, London, Glasgow, Berlin and Paris. Later, railway owned hotels and some commercial Department Stores like Simpsons and Eatons had a telegram booth set up to forward messages.

Telegraphic and telephone service via radio transmission was even available on Canadian National's 'The International Limited' and for a short time on 'The Maple Leaf.'

Signs or posters were created to draw the attention to telegraphic services available at a particular location. At railway stations, signs were generally located in an obvious setting. The signs were commonly found perpendicular to the bay window and just above eye level as one entered the Waiting Room. Other locations, especially the CNR, CNoR and Grand Trunk, placed their station telegraph signs flush against the building facing the hitching post or parking lot areas and flat against the building at the Waiting Room entrance. Letters used on the signs were upper case block style. Usually the parent railway company name was on the sign with 'Telegraph' much larger. Stations posted a variety of styles and simple designs in one form or another, however, there were no fixed rules (see Appendix Eight). Portable stations were rarely posted with a telegraph sign as they were chiefly involved in railway operations. Typically, stations outside city limits had plain signs generally with white painted block letters on a black background - although there were variations as Company logos changed. For example, the CNR did have some telegraph signs designed with the new Maple Leaf logo and even a few 'noodle' ones did appear.

The telegraph company advertised their telegram service. The railways used short, concise, direct and sometimes eye catching graphics. Direct advertising was done in the public railway timetable. Newspaper ads were not common although they did appear.

A closeup of the CNR Ticket-Telegraph office in Montreal. These offices served as travel agents, arranging and confirming local to around-the-world trips by rail and sea. They also sold local and nationwide rail tickets and of course, telegrams. Office front windows were often like the Christmas windows of Eaton's and Simpson's, except year round. Posters of far off places and advertisements were always enticing. Animated displays were common generally around Christmas and Easter.
(CNR Public Affairs, #19982 and 8226-2, Footte)

An example of the ticket office and telegram office. Note the posters and services available from both offices. This photo was taken at Prince Rupert, British Columbia.

(CNR Public Affairs, #19952)

MONTREAL TELEGRAPH CO.

HEAD OFFICE, MONTREAL,

CAPITAL - - - - - - $1,000,000.

HUGH ALLAN, PRESIDENT; SIR W. E. LOGAN, GEORGE W. CAMPBELL, M. D., PETER REDPATH AND ANDREW ALLAN, DIRECTORS; JAMES DAKERS, SECRETARY.

THIS COMPANY was organized in January, 1847, with a capital of $60,000. The line extended from Toronto to Quebec, a distance of 540 miles; nine offices were opened and thirty-five persons employed. The number of messages transmitted during the first year was 33,000.

Since then the Company has, year by year, continued to extend its lines, not only in the Provinces of Ontario and Quebec, where nearly every village of any importance has been afforded ample telegraphic facilities, but also through the Province of New Brunswick, and in the States of Maine, New Hampshire, Vermont, New York and Michigan, in which 110 offices are owned by this Company.

The statistics up to 30th November, 1870, show that the Company possesses 7,800 miles of poles, 12,147 miles of wire, 640 offices and 996 employees. Number of commercial messages sent over the lines during the year 1870, was 1,060,000. Number of words transmitted over Atlantic cable 153,092. Number of words furnished to the press upwards of 8,000,000. To give an idea of the facilities that the Company has at its command it is only necessary to mention that during the last Parliamentary Session it was customary to transmit night after night, from its office at Ottawa, 20,000 to 50,000 words to the press, and on the night of 17th February, as many as 67,251 words were transmitted.

The lines of the Company begin at Sackville, N. B., and extend to Sarnia at the outflow of Lake Huron.

They also run from the United States border to the Georgian Bay and the most northern towns of Canada.

The Company has wires along all the Railways in the Provinces of Ontario and Quebec, and as soon as the Intercolonial Railway is built their wires will be extended over the whole of that road.

In transmitting cable despatches they are sent direct to Sackville, N. B., the present terminus of the Company's lines, over a circuit of 750 miles.

The Company has likewise branches from Montreal working direct through its connections to the following cities in the United States:—

Portland, Maine	300 miles	Oswego, N. Y.	300 miles
Boston, Mass	350 "	Buffalo, N. Y.	450 "
New York, N. Y.	450 "	Detroit, Mich.	550 "

During the year 1870 the Company erected 821 miles of poles, strung up 1,920 miles of wire, and opened 96 new stations. In 1871 they will string an additional wire of 450 miles from Montreal to Buffalo, and have agreed to build lines through the remote districts of Bonaventure and Gaspe, which will be of great service to the fishing interest in that neighborhood, as well as to the shipping entering the St. Lawrence.

Tariff—25 Cents for 10 Words and 1 Cent for Additional Words.

Canadian Pacific Telegraphs

The telegraph system of the C.P.R. not only extends along the entire length of the railway, but also reaches every point of importance off the line of railway in the Dominion of Canada.

The **COMMERCIAL CABLE CO.** (Mackay-Bennett system) gives the **C.P.R.** the most direct connection with Europe, Asia and Africa.

Also in direct connection with the **BRITISH PACIFIC CABLE** to Australia, New Zealand, etc., and the **COMMERCIAL PACIFIC CABLE** to the Hawaiian Islands, Philippine Islands, China and Japan.

The **POSTAL TELEGRAPH CO.**, of New York and San Francisco, enables the **C.P.R.** to reach all important points in the United States.

The **HALIFAX-BERMUDA & DIRECT WEST INDIES CABLE CO.** gives connection with Bermuda, Jamaica and points in the West Indies.

Direct connection with the Dominion Government Telegraph Line to the Yukon and points in Alaska.

A tariff of charges for Telegrams is posted at all Telegraph Offices, and is kept by all Sleeping and Parlor Car Porters for inspection. Senders of messages are requested to compare the charges made with the rates on tariff.

To ensure quick dispatch to all points, see that your telegrams are written on **C.P.R.** Telegraph Blanks and are handed in at C.P.R. Offices.

JAMES KENT,
MANAGER TELEGRAPHS.

HEAD OFFICE, MONTREAL.

Telegram Ads from several Canadian public timetables. Though not always prominent, they could be found several times throughout a timetable and as one line dividers between timetable columns, and as headers or footers just before page numbers. The earliest ones were 'long-winded' explanations bestowing the many benefits of sending telegrams; later they became very concise and graphic.

(R.G. Burnet)

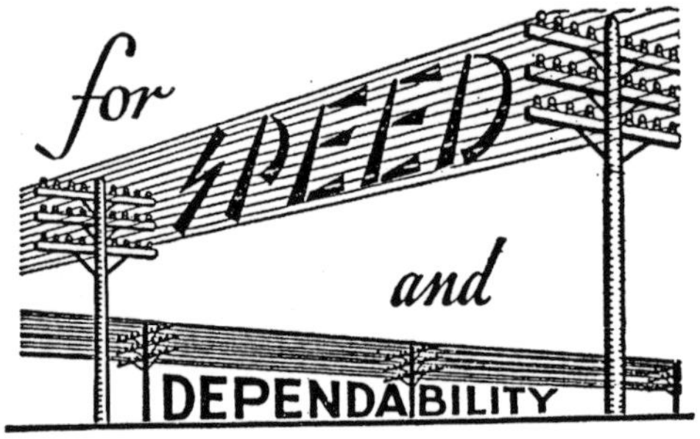

TELEGRAPH, CABLE and WIRELESS SERVICE
TO ALL PARTS OF THE WORLD

Money Transferred by Telegraph—
Safely and Promptly

Avoid Disappointment—Reserve Hotel or other accommodation by wire. Sleeping or Parlor Car conductors will arrange for despatch of telegrams if desired.

NEW . . . FAST
EVENING AND SUNDAY
TELEGRAPH SERVICE

AT GREATLY REDUCED RATES

Messages will be accepted at any time, for transmission and delivery immediately after 7.00 p.m., or at any time Sunday.

Reduced rates between offices in Canada—40 words for the price of a 10-word "Day Message"—50% reduction for immediate prepaid reply!

EXAMPLES OF SAVINGS—40-WORD MESSAGE

	NEW RATE		OLD RATE	
	One Way	Ppd. reply (If desired)	One Way	Ppd. reply (If desired)
TORONTO-WINNIPEG	$ .80	$ .40	$2.30	$2.30
TORONTO-VANCOUVER	1.25	.63	3.65	3.65
TORONTO-MONTREAL	.40	.20	1.30	1.30

TELEGRAMS ACCEPTED BY TELEPHONE

CANADIAN PACIFIC TELEGRAPHS

FAST TELEGRAMS!

Are Vital Communications

Command Preferred Attention

Speed Up Production

Get Results

When Time Is a Factor

use

FAST TELEGRAMS!

WIRE FOR IT Save Time!
Smooth-running business depends on swift communication. To speed commerce, Canadian National Telegraphs offer far-flung facilities. Avoid misunderstanding . . . ensure prompt action and permanent written records—with telegrams.
CANADIAN NATIONAL TELEGRAPHS

PRIVATE WIRE Teletype SERVICE
. . . provides accurate, continuous liaison with all branches of your organization, near or far. It's fast, versatile . . . and you'll be surprised at the economy!
Ask your local telegraph office for details.
CANADIAN NATIONAL TELEGRAPHS

Friendly, personal way to wire your greetings
ON BIRTHDAYS
ANNIVERSARIES
WEDDINGS
SPECIAL OCCASIONS
CANADIAN NATIONAL
"Social Telegrams"
—at any CN Telegraph Office

WIRE . . . and let them know you're coming
CANADIAN PACIFIC TELEGRAPH OFFICE
on platform at all principal stops

Telegram speed, without delay, was demanded. Technological advances such as the Stop-Start printer, Telex, high speed dedicated circuits, and carrier current greatly improved the transmission of the telegram. The courses, strictly run by the railway and developed by railway officials, produced highly skilled operators. These operators could double and triple commercial and railway message transmissions. These elite operators were known as "Recording Operators." Recording operators, known colloquially as 'lightning ops,' handled multitudes of telegrams hourly and without transmission delays. The vibroflex 'bug' was the preferred key used to transmit telegrams by these operators. Having direct connections was essential, especially when transferring messages to different companies. Canadian Pacific had their own direct worldwide telegraphic connections by the 1920s, along with many other connections to different companies. Canadian National, by the late 1940s, was able to proclaim, "... DIRECT TELEGRAPH SERVICE TO 21,316 PLACES IN CANADA AND THE UNITED STATES WITHOUT DELAYING THE TELEGRAMS IN ROUTE BY TRANSFERRING THEM TO ANOTHER COMPANY FOR ONWARD TRANSMISSION [sic]."[4]

Handling telegrams was another aspect. When a telegram was received, it had to be dealt with promptly. And delay in delivery or transmission was not tolerated. On arrival of a telegram (and similarly with waybills), consignees were advised of their shipment. Agents were advised to keep accurate records of whom they contacted, arrival time and train number, when they notified a customer, condition of the goods and so on. It was a precise way to avoid unnecessary insurance claims. Rural agents were known to drop off telegrams and packages personally or even hire a local student to deliver them. In the city there was the long line of bikes in the bicycle racks and motorcycles parked in front of the CNR Telegram office on Front Street or the National Building at 347 Bay Street. It was a great summer job for students delivering telegrams for CNT. The competition for these positions was intense, and if one got the job, one would be delivering messages all over the city and making some fine tips. During World War II, the CPR reported: "Not all the masculine duties, however, deprive the girls their right to look glamorous. Attractive young misses, in smartly designed uniforms, are now replacing the company's telegraph boys in several centres. They have proven themselves just as expert on their bikes; just as speedy, and every bit as efficient in discharging their duties."[5]

Uniformed Telegraph messengers on bicycles or motorbikes were quite common for many years in all Canadian cities and towns. This photo was taken about 1950 outside the National Building at 347 Bay Street, Toronto.

(CNR Public Affairs, #X32088)

Two fundamental procedures existed for agents when handling telegrams in a Morse operated office:

> Commercial messages must be transmitted to the assigned relay point even if the office of the destination is on the same circuit. In special cases, the destination office on the same circuit may take a copy to eliminate double transmission and save wire space. At offices where reception is made on a typewriter, use the double receiving Blank Form 6123 (bilingual Form 6123B is available for use when required). A carbon copy of the message must be made.
>
> Where the reception is in long hand, agents and operators will copy incoming telegrams on receiving Blank Form 6121 (bilingual Form 6121B is available where required) making a carbon copy. Commercial telegrams must not be copied on sending blanks or on Railway Service forms. It should not be necessary to recopy a telegram for delivery. Message of a social nature, congratulations, best wishes etc., should be copied neatly on the social receiving Blank Form 6126 and enclosed in a social envelope, Form 6134.6

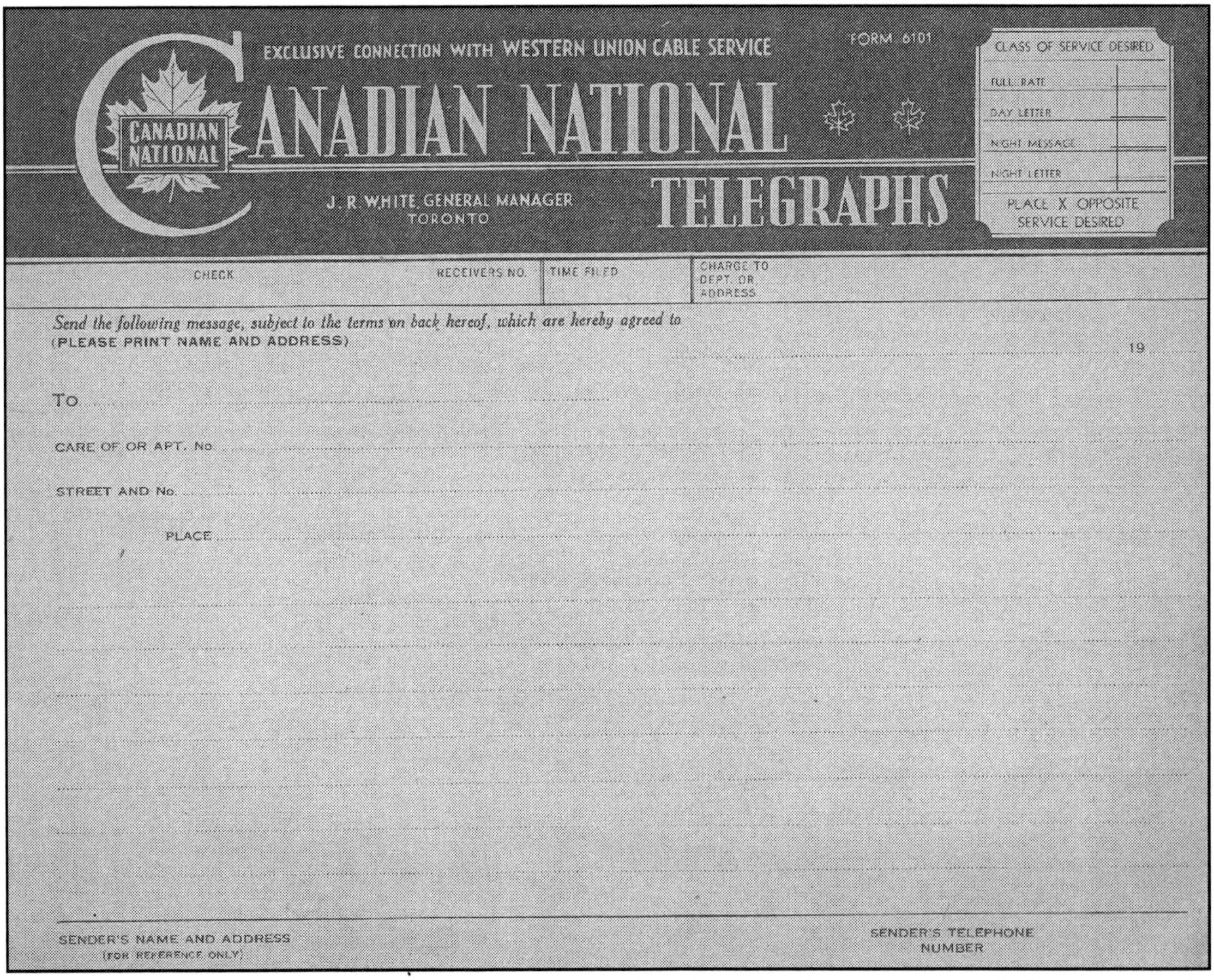
EXCLUSIVE CONNECTION WITH WESTERN UNION CABLE SERVICE

FORM 6101

CANADIAN NATIONAL

CANADIAN NATIONAL TELEGRAPHS

J. R. WHITE, GENERAL MANAGER
TORONTO

CLASS OF SERVICE DESIRED	
FULL RATE	
DAY LETTER	
NIGHT MESSAGE	
NIGHT LETTER	

PLACE X OPPOSITE SERVICE DESIRED

CHECK	RECEIVERS NO.	TIME FILED	CHARGE TO DEPT. OR ADDRESS

Send the following message, subject to the terms on back hereof, which are hereby agreed to
(PLEASE PRINT NAME AND ADDRESS)

19

TO

CARE OF OR APT. No.

STREET AND No.

PLACE

SENDER'S NAME AND ADDRESS
(FOR REFERENCE ONLY)

SENDER'S TELEPHONE NUMBER

CNR telegram blank Form 6101 - A standard fill in form for the customer and the reverse side providing instructions and legal obligations.
(R.G. Burnet)

CANADIAN NATIONAL TELEGRAPH COMPANY

(operating its own lines and those of the Great North Western Telegraph Company, the Grand Trunk Pacific Telegraph Company and Canadian Government Railways).

HEREINAFTER CALLED THE COMPANY.

TERMS AND CONDITIONS UPON WHICH TELEGRAPH AND CABLE MESSAGES SHALL BE TRANSMITTED ARE PRESCRIBED BY ORDER NO. 49274, DATED DECEMBER 5TH, 1932, OF THE BOARD OF TRANSPORT COMMISSIONERS FOR CANADA AND PUBLISHED IN THE CANADA GAZETTE.

It is agreed between the sender of the message on the face of this form and this Company that said Company shall not be liable for damages arising from failure to transmit or deliver, or for any error in the transmission or delivery of any unrepeated telegram, whether happening from negligence of its servants or otherwise, or for delays from interruptions in the working of its lines, for errors in cipher or obscure messages, or for errors from illegible writing, beyond the amount received for sending the same.

To guard against errors, the Company will repeat back any telegram for an extra payment of one-half the regular rate; and, in that case, the Company shall be liable for damages suffered by the sender to an extent not exceeding $200.00, due to the negligence of the Company in the transmission or delivery of the telegram.

Correctness in the transmission and delivery of messages can be insured by contract in writing, stating agreed amount of risk, and payment of premium thereon at the following rates, in addition to the usual charge for repeated messages, viz.: one per cent for any distance not exceeding 1,000 miles, and two per cent for any greater distance.

This Company shall not be liable for the act or omission of any other Company, but will endeavour to forward the telegram by any other Telegraph Company necessary to reaching its destination, but only as the agent of the sender and without liability therefor. The Company shall not be responsible for messages until the same are presented and accepted at one of its transmitting offices; if a message is sent to such office by one of the Company's messengers, he acts for that purpose as the sender's agent; if by telephone, the person receiving the message acts therein as agent of the sender, being authorized to assent to these conditions for the sender. This Company shall not be liable in any case for damages, unless the same be claimed, in writing, within sixty days after receipt of the telegram for transmission.

No employee of the Company shall vary the foregoing.

CLASSES OF SERVICE

FULL RATE TELEGRAM

A full-rate expedited service.

DAY LETTER

A deferred day service at rates lower than the standard telegram rates as follows: One and one-half times the ten-word day message rate for the transmission of 50 words or less, and one-fifth of the initial rate for such 50 words for each additional 10 words or less.

Day letters may be forwarded by the Company as a deferred service, and the transmission and delivery of such Day Letters are, in all respects, subordinate to the priority of transmission and delivery of full-rate messages.

Day Letters may be delivered by the Company by telephoning the same to the addressees, and such deliveries shall be a complete discharge of the obligation of the Company to deliver.

Day Letters are received subject to the express understanding and agreement that the Company does not undertake that a Day Letter shall be delivered on the day of its date absolutely and at all events, but that the Company's obligation in this respect is subject to the condition that there shall remain sufficient time for the transmission and delivery of such Day Letter on the day of its date during regular office hours, subject to the priority of the transmission of full-rate messages under the conditions named above.

NIGHT MESSAGE

Accepted up to 2 a.m. at reduced rates, to be sent during the night and delivered on the morning of the next day after their date. At places where the Company's offices are not open on Sundays, delivery will be made on the morning of the next ensuing business day.

NIGHT LETTER

Accepted up to 2 a.m. at reduced rates, to be sent during the night and delivered on the morning of the next day after their date. At places where the Company's offices are not open on Sundays, delivery will be made on the morning of the next ensuing business day. The rates for Night Letters are still lower than the standard Night Message rates, as follows: The standard day rate for 10 words for the transmission of 50 words or less and one-fifth of the initial rate for such 50 words for each additional 10 words or less. The minimum charge for transmission over the Company's lines of any Night Letter will be 30 cents.

Night Letters may, at the option of the Company, be mailed at destination to the addressees, and the Company shall be deemed to have discharged its obligation in such cases with respect to delivery by mailing such Night Letters at destination, postage prepaid.

Further, the Grand Trunk demanded that "Agents must practice and enforce economy in the use of supplies. The use of the pencil on business documents or papers is strictly forbidden. Pen and ink must be used."[7] An established GTR rule demanded secrecy for all telegrams and their contents by telegraph staff: "Telegrams addressed to officials must be put in an envelope, sealed and properly addressed in each and every case. The contents of all telegrams must be held as confidential." This was standard practice with all railway companies.

Commercial operators were paid a set salary and commissions were possible. They were considered the most select telegraphers, even over recording operators, because they only transmitted or received telegram messages. Terminal offices offered pay cheque incentives of one or two cents more if sixty plus messages were handled per hour. The commission rate did improve to 10 and 15 percent. Railway agents on salary (Figure One) could earn commissions for each message as well.

FIGURE ONE:
Examples of wages earned by an Agent or Operator between Montreal and Toronto.

STATION NAME	POSITION	WAGE PER MONTH	HOUSING	CONTRACT DATE
Mallorytown	Agent	$133.00	Free HFL*	June 16,1927
	Operator	$122.00	Roomer or supply own	"
Port Union	Agent	$137.00	supply own	"
	Operator	$122.00	supply own	"
Danforth	1st Operator	$148.00	supply own	"
	2nd Operator	$147.00	supply own	"
	3rd Operator	$150.00	supply own	"
Mallorytown	closed			Dec. 1,1947
Port Union	Agent	$190.00	supply own	"
	Operator	$175.00	supply own	"
Danforth	1st Operator	$201.00	supply own	"
	2nd Operator	$202.00	supply own	"
	3rd Operator	$205.00	supply own	"

* The Grand Trunk Railway supplied a house for the Agent and his family in several locations. The houses were categorized as 'Free HFL' or Free House Fuel and Lights. After amalgamation in 1921, Canadian National eliminated most of these facilities. A complete list may be found in various contract editions of the CNR Rules and Wages for Telegraphers.

During World War Two, the telegraph office was a sensitive place. Although telegraph agents and operators were exempt from military service, the CNR had over 14,000 employees go overseas, and many were telegraphers. Many women telegraphers became extensively involved during the war.

For those men and women in the station or terminal office, they were reminded of how easily the code could be tapped into by the enemy. Security was more intense in some locations and less in others. For example, it was commonplace for the CPR and CNR to send a telegraph employee with possible connections to the enemy, into a less sensitive location. Decoder books to code/decode military messages were used. However, it was common practice for most railway employees to already know about troop movements: although a message might be sent in code, the same message was relayed in English, to the Sleeping and Dining Car Department so meal and blanket requisitions could be filled for the troops. Agents, nonetheless, in the bigger terminal offices were advised to be watchful for anything out of the ordinary. Further, Canadian National President R.C. Vaughan demanded discretion by all personnel, as did other railway employers. (Canadian National Telegrams handled over 55 million telegrams between 1939 and 1945, not including military messages.)

CANADIAN NATIONAL RAILWAYS

R. C. VAUGHAN,
PRESIDENT.

MONTREAL, July 13th, 1942.

TO ALL OFFICERS AND EMPLOYEES:

Notwithstanding repeated governmental warnings issued through the press, by posters and by other means, vital information relating to Canada's war effort and that of her allies is still being publicly discussed by careless and thoughtless citizens and may possibly be reaching enemy hands.

Under the circumstances, it is felt that all officers and employees should be further cautioned as to the necessity of maintaining absolute secrecy respecting the movement of ships, military trains, munitions, equipment and supplies, harbor and shipyard facilities, munition plants and all other activities or movements connected with the war effort, past, present, or future.

Information concerning these matters acquired in the course of duty must not be imparted to persons other than those who for the proper conduct of the Company's business are entitled and required to receive it. It should not and must not be communicated to members of employees' families, or to close friends or relatives. Officers and employees whose duties bring them into contact with the public are particularly warned against being drawn into conversations involving a discussion of the facilities and functions of the Canadian National Railways and their subsidiaries and Canadian National Steamships, in so far as they are in any manner connected with the war effort.

The average citizen is in no position to judge whether the discussion of information pertaining to the war effort will eventually assist the enemy. The cold fact is that these matters are military secrets and must be respected as such.

The unauthorized disclosure of information which is prejudicial to the safety and interests of the state comes within the purview of the Official Secrets Act and renders the offender liable to fine or imprisonment, with or without hard labor, or both.

The enemy agent is anxious to obtain the information which you are hereby advised to keep secret; he has sharp eyes and an alert brain. This warning, therefore, carries with it the instruction to effectively and continually safeguard written instructions, documents and records which in any way have a bearing upon the war effort of the United Nations.

President.

From R.C. Vaughan, President of Canadian National Railways during World War Two. These notices were sent to all agents and railway employees. (R.G. Burnet)

CANADIAN NATIONAL RAILWAYS

MONTREAL, QUE., JUNE 8TH, 1943.

TO ALL OFFICERS AND EMPLOYEES:

The military authorities are greatly worried by the amount of accurate information of value to the enemy which is known to the general public and they are asking us to help them to stop the spread of such information which may result in a grave loss of life.

For example, the Royal Canadian Air Force have been astounded at the extent and accuracy of information about transport and troop train movements in circulation prior to steamship sailings from Canadian ports.

It is not suggested that railway men are responsible for making such information known. But it is felt that because many of us on the railways must be given advance notice of such movements and because of our close, personal contact with the armed forces generally, we should be particularly careful not to tell what we know and not to be drawn into talk dealing with troop movements or overseas shipments of any kind.

Quite apart from the penalties involved for careless talking, it is our plain duty as loyal citizens to keep what we know to ourselves and to encourage all others to do likewise.

In this vital contribution to the success of our arms and the welfare of our nation, I bespeak the whole-hearted assistance of every man and woman on the Canadian National System.

President.

From R.C. Vaughan, President of Canadian National Railways during World War Two. These notices were sent to all agents and railway employees. (R.G. Burnet)

The most difficult telegram to handle was a death notice. The railway companies demanded that agents receiving a death or Missing In Action telegram give it special care. Serious wounds or injuries were left to the agent's discretion. "Delivery must be made in written form between 7 a.m. and 10 p.m. It must be made under the personal supervision of the agent to lessen the shock. Exercise of initiative and judgement is necessary."[8] To assist the agent with notifying next of kin, he/she often examined the surname. It was generally known that if the last name was Smith, they were likely Anglican; if MacIntrye, Scottish and Presbyterian, or O'Reilly, Irish and Catholic. The agent would contact the local minister or priest and inform them of the news. A church leader, to the relief of many agents, would deliver the telegram.[9]

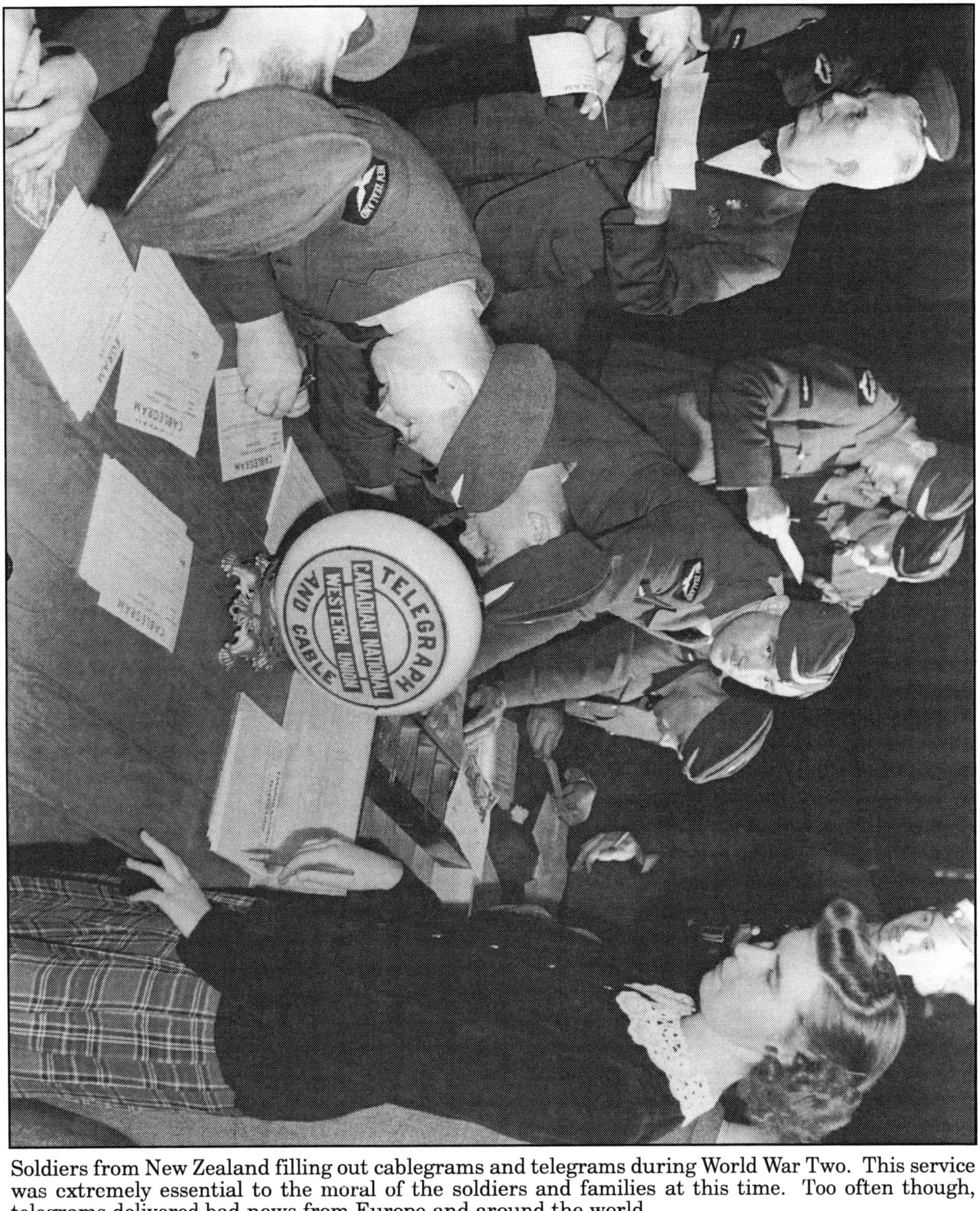

Soldiers from New Zealand filling out cablegrams and telegrams during World War Two. This service was extremely essential to the moral of the soldiers and families at this time. Too often though, telegrams delivered bad news from Europe and around the world.
(CNR Public Affairs, #X14974)

The telegraph was the major communication tool for the railway and was not to be abused or misused by employees. Notices were sent around as reminders. The monthly Educational Demerit Sheets was another way to highlight employee transgressions or merits. Therefore, unless the circumstances were exceptional, "... employees were not to use the circuit. If supplies were required, an agent had to use the Railway Correspondence form and ship it by train.... The use of the telegraph wires is to be avoided excepting in cases of absolute necessity when there is not sufficient time to order [tickets] by mail. Telegrams must be signed in the name of the agent, viz: 'John Jones,' and not 'Agent.' Requisitions should be followed by mail, endorsed 'Confirmation of Telegram.'"[10]

Early folk-lore emerged when the telegraph was still an unknown entity. Some of these tales included:

- people were disgusted to learn that bundles of laundry or mail could not be sent home for washing or sorting over telegraph wires;
- the telegraph was a tool for gamblers who would eventually wreck the status quo of industry, the economy and social life generally;
- the telegraph would be used to overthrow the government;
- apparently an agent snuck in on his wife late one night after being out with the boys; his boots were extremely dirty so he hung them up on the telegraph wire; everyone who walked by the next day was told that his boots had come by night telegram - and he was believed.

Dreams about telegrams have turned up in dream analysis books between 1880 and 1900. Five major themes can be shown if one dreamt about a telegram: 1) To receive a telegram meant something bad would happen to you, such as a friend talking behind your back and ruining your good name; (2) to dream of sending a telegram meant some bad business dealings or a friend might estrange you; (3) to dream of being in the telegraph office or station meant one would have an unfortunate engagement, unfortunate love affairs, or a failed marriage; (4) if one dreamed of being the operator sending the telegram, one would be affected only if the person receiving the telegram was involved; (5) if one was hapless enough to be the telegraph operator sending bad news in a dream, he would be subject to and affected by the gossip that followed.

Because the station was often a gathering place, "In [the] early days, it was customary for the towns' people to gather about the [telegraph] office to secure any news coming over the wire. In fact, if there were items of importance, a bulletin would be displayed outside."[11]

Finally, the article below appeared in the Toronto Star 3 June 1905:

INSANITY ON THE INCREASE
Dr. Sheard Says It Has Increased
20 Per Cent. in Ten Years.

THE RESULT OF CITY LIFE

"Insanity has increased 20 per cent in this city in the past ten years," said Dr. Sheard, Medical Health Officer, today.

"What caused the increase?" he was asked.

"The changed conditions of living," was the reply.

"The flocking to the cities, where there is so much excitement, and where everything proceeds at high pressure.

"The telephone and telegraph are responsible. Business men have no relaxation now from business. Nowadays when a business man goes home he is recalled to business complexities by phone and telegraph messages."[12]

[The article goes on to also blame stress and drug abuse.]

II.
TELEGRAM TYPES and SERVICES

Originally two types of commercial telegrams were available: the full rate message and night message. The full rate telegram was determined by the individual company and was designed to be competitive. A reduced rate for the night message was also set by the individual company. Canadian Pacific introduced a night letter and a day lettergram about 1890. Around 1900, the Great North Western Telegraph Company initiated a full package of services to commercial customers. The GNWTCo introduced a Day Message [code:DM, no symbol used], Day Letter [code:DL or BLUE], Night Message [code:NM or NITE] and Night Letters [code:NL] and each category had a different charge. During and towards the end of the Depression, a flurry of telegraphic services and cheaper rates began appearing. Because of sophisticated technology like Carrier Current and Telex, Weekend Letter [code:WLT] and Weekend Message [code:WMT], Urgent Message and Special Occasion telegrams became commonplace.

Around 1936 Canadian Pacific Telegraphs introduced new rates and services. The service was classified under Message Transmission Service. Canadian Pacific stated: "The department is endeavouring to meet the communications needs of the public, as progress and changing conditions require, by improvements in existing services and by providing new services."[13] One new service involved a discount for advertising agencies and newspapers from 15 to 50 per cent deducted from the charged tariff based on the volume sent: the more sent, the more discount. The seasonal telegram or "Yuletide" appeared after World War One as did the Special Occasions telegrams for birthdays, New Years, Congratulations, Easter, Anniversaries and so on. Each special occasion had a fixed text message length, rate, special paper and envelope used. Interestingly, the special forms were not always available in the more remote regions. Canadian Pacific Telegraphs also introduced the Post Lettergram, developed to compete directly with Royal Mail, later Canada Post. Canadian Pacific operated a money order service with Canadian Pacific Express where "... money is transmitted by telegraph and cable to practically all parts of the world."[14] Canadian National offered similar and competitive services.

Masthead from a Yuletide telegram of 1921-1922 from Canadian Pacific. These were special forms with a fixed message and tariff. This one was in the usual Christmas colours.
(R.G. Burnet)

An example of a special CNR Greetings telegram Form 6126-B. This type had a fixed tariff and message with bright colours.
(CNR Public Affairs, #X42017)

The Singing Telegram originated with George P Oslin at Western Union. The earliest recorded date was 28 July 1933. It later received world wide exposure and was enjoyed by most recipients. As an example, the following note was attached to a telegram with a special request for the messenger to follow through with: "Instead of having the boy sing this message to him, please have the messenger slap his face."[15]

Because of World War Two, nonessential telegram services were either reduced or eliminated. Canadian National continued to handle "... as promptly as conditions permit ... [however], certain non-essential services may of necessity be curtailed or withdrawn."[16] The abolished services included the popular Singing Telegram, all Special Occasion Messages, and fast on time delivery could no longer be guaranteed due to the volume of government services, the military and railway business needs.

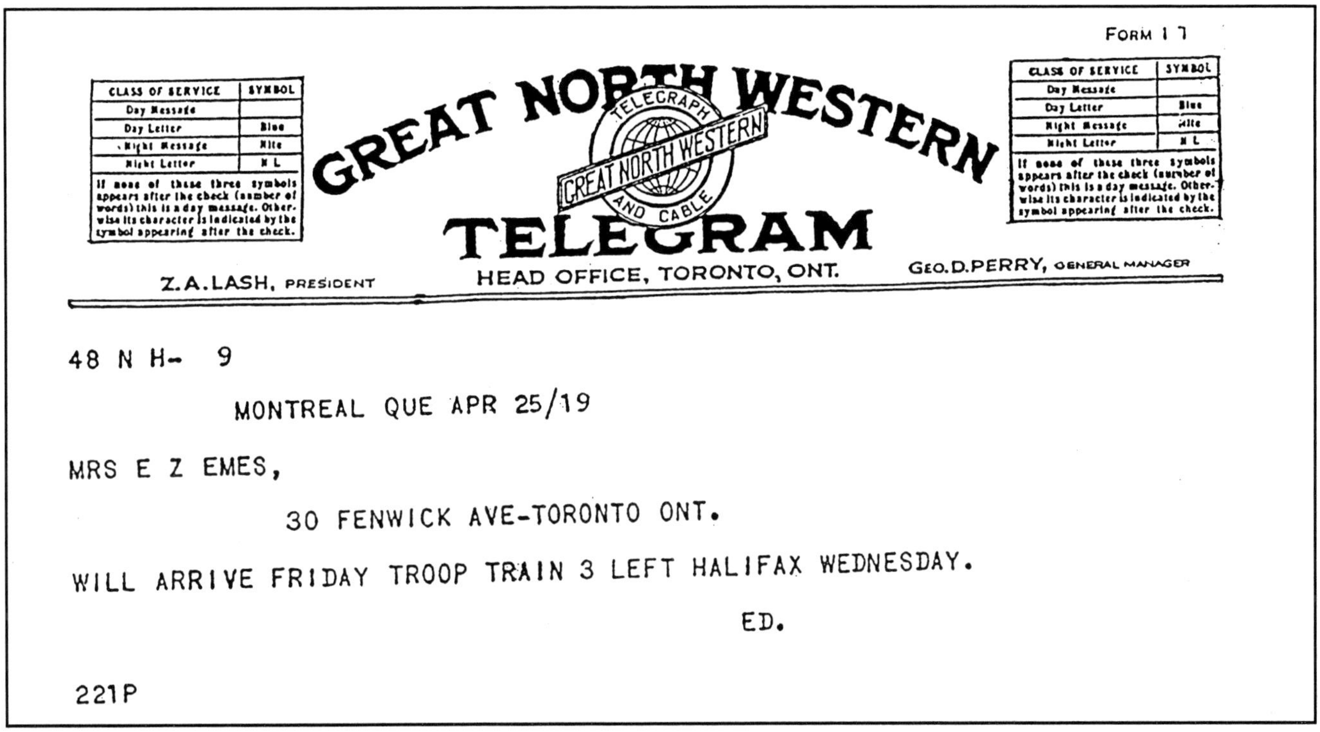

FORM 17

CLASS OF SERVICE	SYMBOL
Day Message	
Day Letter	Blue
Night Message	Nite
Night Letter	N L

If none of these three symbols appears after the check (number of words) this is a day message. Otherwise its character is indicated by the symbol appearing after the check.

GREAT NORTH WESTERN TELEGRAPH AND CABLE

GREAT NORTH WESTERN TELEGRAM

HEAD OFFICE, TORONTO, ONT.

Z. A. LASH, PRESIDENT — GEO. D. PERRY, GENERAL MANAGER

48 N H- 9

MONTREAL QUE APR 25/19

MRS E Z EMES,

30 FENWICK AVE-TORONTO ONT.

WILL ARRIVE FRIDAY TROOP TRAIN 3 LEFT HALIFAX WEDNESDAY.

ED.

221P

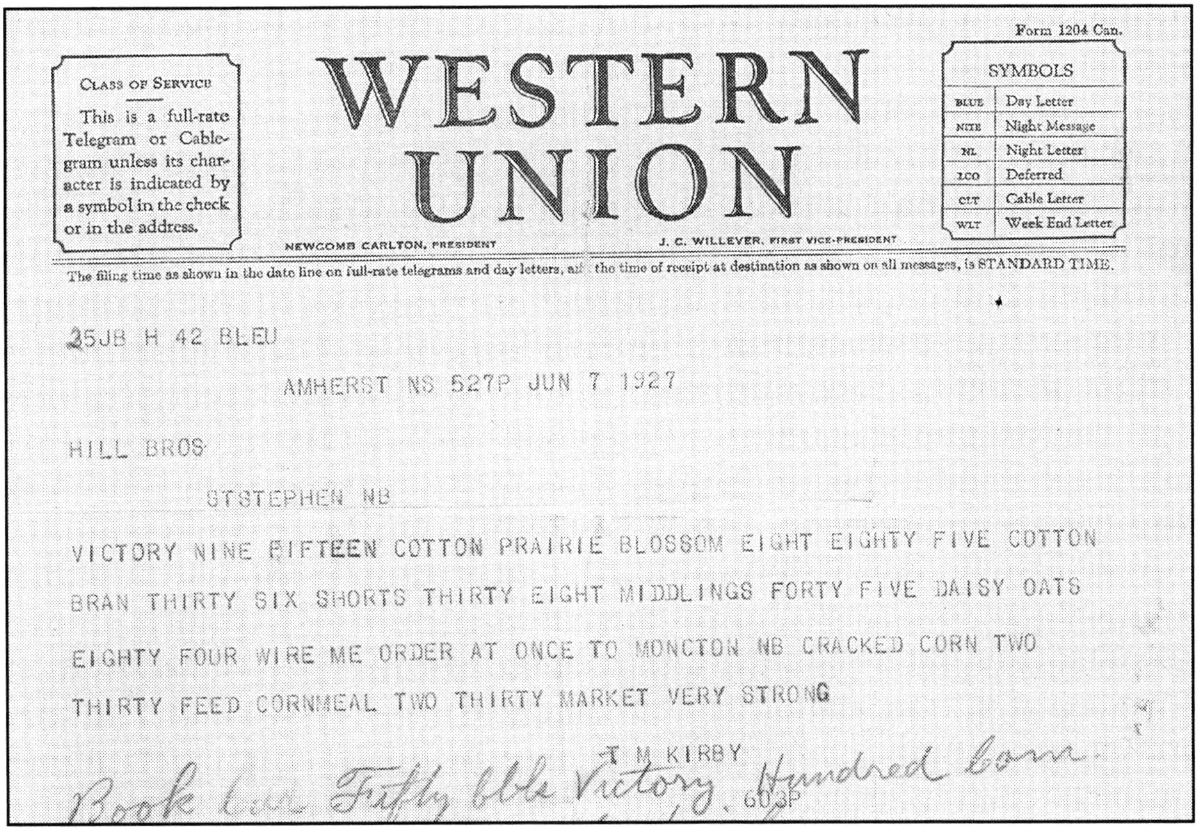

Form 1204 Can.

CLASS OF SERVICE

This is a full-rate Telegram or Cablegram unless its character is indicated by a symbol in the check or in the address.

WESTERN UNION

NEWCOMB CARLTON, PRESIDENT — J. C. WILLEVER, FIRST VICE-PRESIDENT

SYMBOLS	
BLUE	Day Letter
NITE	Night Message
NL	Night Letter
LCO	Deferred
CLT	Cable Letter
WLT	Week End Letter

The filing time as shown in the date line on full-rate telegrams and day letters, and the time of receipt at destination as shown on all messages, is STANDARD TIME.

35JB H 42 BLEU

AMHERST NS 527P JUN 7 1927

HILL BROS

STSTEPHEN NB

VICTORY NINE FIFTEEN COTTON PRAIRIE BLOSSOM EIGHT EIGHTY FIVE COTTON BRAN THIRTY SIX SHORTS THIRTY EIGHT MIDDLINGS FORTY FIVE DAISY OATS EIGHTY FOUR WIRE ME ORDER AT ONCE TO MONCTON NB CRACKED CORN TWO THIRTY FEED CORNMEAL TWO THIRTY MARKET VERY STRONG

T M KIRBY

Book Car Fifty bbls Victory Hundred Corn.

603P

Samples of various graphic telegram mastheads and personal messages. Note the rubber stamp indications where the message was received or sent from to the family. (R.G. Burnet)

C.N.T. 6123.

CLASS OF SERVICE	SYMBOL
Full-Rate Message	
Day Letter	D L
Night Message	N M
Night Letter	N L

If none of these three symbols appears after the check (number of words) this is a full-rate message. Otherwise its character is indicated by the symbol appearing after the check.

CANADIAN NATIONAL TELEGRAM

TELEGRAPH CANADIAN NATIONAL AND CABLE

411P

Exclusive Connection with WESTERN UNION TELEGRAPH CO. Cable Service to all the World Money Transferred by Telegraph

D. E. GALLOWAY, ASSISTANT VICE-PRESIDENT, TORONTO, ONT.

STANDARD TIME

RAD54 26=SOUTHERN PINES NCAR 14 305P

MRS WILLIAM J HAWN=

:BRACEBRIDGE ONT=

WE ARE NOT IN SOUTHERN PINES SNOWING HARD IN SUNNY SOUTH
WE HAVE JUST RECOVERED FROM FLU WILL WRITE SOON LOVE TO
MOTHER DAD AND GERTRUDE=

:BEA.

468

C.D. 1M

CANADIAN PACIFIC COMMUNICATIONS

CANADIAN PACIFIC TELEGRAPHS

DIRECT CONNECTION WITH
POSTAL TELEGRAPH - CABLE CO.
COMMERCIAL CABLES - - IMPERIAL CABLES

MONEY TRANSFERRED BY TELEGRAPH

This is a full-rate Telegram or Cablegram unless otherwise indicated by signal in the check or in the address.

TELEGRAM		CABLEGRAM	
FULL RATE		FULL RATE	
DAY LETTER	DL	CODE	CDE
NIGHT LETTER	NL	DEFERRED	LC
NIGHT TELEGRAM	NM	NIGHT CABLE LETTER	NLT

W. D. NEIL, GENERAL MANAGER OF COMMUNICATIONS, MONTREAL.

STANDARD TIME

233 S 17 RUSH XU

HALIFAX NS SEPT 28 1936 543p

HILL BROS

STSTEPHEN NB

SUBJECT CONFIRMATION HIGHLOAF SIX ELEVEN BRAN TWENTY SEVEN SIXTY
SHORTS TWENTY NINE SIXTY BETTER BOOK ANSWER QUICK

GIBSON BROS

715p

Book One Fifty High Loaf bal feeds
quote Barley Meal cr Corn

(Printed in Canada) FORM T. D.1X

CANADIAN PACIFIC RAILWAY COMPANY'S TELEGRAPH

TELEGRAM

CABLE CONNECTIONS TO ALL PARTS OF THE WORLD

J. McMILLAN, General Manager of Telegraphs, Montreal.

STANDARD TIME

1272

YO.RA....8

BOISSEVAIN,MAN AUG 15- 4,55 PM

REV & MRS ISON,

CARE W ALDRIDGE , 333 MCGEE ST , WINNIPEG,MAN

CONGRATULATIONS AND BEST WISHES FROM THE GLOVERS BOISSEVAIN .

JENNIE

Canadian Pacific C.D. 1R

TELEGRAM

WORLD WIDE COMMUNICATIONS

"The filing time shown in the date line is STANDARD TIME at place of origin. Time of receipt is STANDARD TIME at place of destination"

RNA523 RS

RECEIVED 4 KING WEST DIAL JA-7-2023

TORONTO ONT 27

BAGGAGE AGENT

1955 APR 27 PM 8 40

HAMILTON ONT

CONFIRMING PHONE CONVERSATION DATE FOR SC OUT OF LOCKER#290 FWD HERE FIRST TRAIN ADVISING. ALSO KEY #290 TO YOU ON TRAIN #329 DATE

G CROMB

OK RB

Railway employees used telegram franks. Franks ensured free conveyance of a telegram by a railway employee or official. The frank began in the late 1880s as one of many conveniences and privileges given to railway employees and families. The frank had a unique individual consecutive counting number, the current company logo, and most often issued yearly.

Frank samples from Canadian Pacific Telegraphs, the Great North Western Telegraph Company, and Canadian National Telegraph.
(R.G. Burnet)

Canadian Pacific Railway issued the first Canadian frank in 1887 and their last frank in 1936. The CPR experimented with different colours but most were black (1902-blue; 1903-brown; 1904-slate). Great North Western Telegraph Company franks were issued between 1890 and 1920. They used colours that followed an irregular pattern. For example, blue was used in 1890, 1897, 1904, 1909, 1917; red in 1891, 1896, 1900, 1905 and 1910; shades of green followed in 1892 (olive green), 1898 (green), 1901 (blue green), 1906 (green) and 1915 (green). Algoma Central issued only two franks in 1890 and 1902. The Canadian Northern Railway printed them between 1904 and 1914 only; the Grand Trunk Pacific 1911 to 1920 only, and Canadian National Railways from 1921 to 1935.

Franks required strict accounting. "A Free Message Report, Form 6185 [1950s], accompanied by the office copies of all sent 'DH Frank' [Dead Head Frank] and 'DH Emp.' [Dead Head Employee] message must be forwarded to the R.A.C. [Telegraph Revenue Accounting Centre] at the close of each month. If no free messages are handled a 'NIL' report is not required by the R.A.C."[17] Today, franks are scarce, especially the booklet of four panes surrounded by thin tissue paper with cardboard covers and a company logo. Of interest, genuinely cancelled franks are extremely rare. Especially desired by collectors are franks still intact on the envelope facing or message with a railway station cancellation.

Revenue statistics from most telegraph companies was placed in the Miscellaneous Revenue or Other Income accounts. The CPR maintained a separate telegraph accounting ledger until 1918 when it was switched to Other Income category. "The dependence of 'Other Income' on the ownership of securities and subsidiary companies, ... and on the reclassification of accounts with reference to the inclusion in later years of net earnings from ... commercial telegraph and [the] news department ... occasioned marked changes in the total."[18]

FIGURE TWO:

Commercial Telegraph Revenue, CPR 1886 to 1918
(Complied from C.P.R. Reports)[19]

Year	Revenue	Year	Revenue
1886*	$ 50,619.28	1904	$ 590,722.89
1887	180,794.89	1905	635,432.50
1888	207,952.76	1906	728,762.24
1889	241,757.76	1907	882,640.87
1890	288,697.36	1908	887,534.41
1891	328,707.60	1909	933,137.61
1892	345,986.39	1910	1,057,783.35
1893	335,736.80	1911	1,196,899.08
1894	342,411.52	1912	1,435,944.28
1895	302,261.34	1913	1,691,953.38
1896	336,030.71	1914	1,613,687.64
1897	377,576.69	1915	1,382,507.16
1898	457,097.36	1916	1,339,161.02
1899	489,807.64	1916 (six months)	853,353.89
1900	not given	1917*	1,770,450.00
1901~	449,058.69	1918	2,852,416.30
1902	492,902.88		
1903	547,438.15		

* Calendar year ~ Fiscal year (six months omitted)

Telegrams have served people for over 150 years. The life of a business, the social life of a community, or the love for an individual is shown in these messages. Telegrams were faster than mail, more reliable, inexpensive and personal. Because of human involvement, telegrams were endeared long after they were received. Like the melody of music recalling an incident from one's past, finding an old telegram in one's cedar chest brings memories to the forefront offering one a chance to reflect whether pleasant or melancholy. The following poem combines the summative mystic and sensitivity of the telegrapher and his youthful customer in a wonderful graphic manner:

THE TELEGRAM

"Is this the tel'graph office?"
asked a childish voice one day.
As I noted the click of my instrument
With its message from far away;
As it ceased, I turned; at my elbow
Stood the merest scrap of a boy,
Whose childish face was all aglow
With the light of hidden joy.

The Golden curls on his forehead
Shaded eyes of deepest blue.
As if a bit of the summer sky
Had lost in them its hue;
They scanned my office rapidly
From ceiling down to floor,
Then turned on mine their eager gaze,
As he asked the question o'er.

"Is this the tel'graph office?"
"It is, my little man,"
I said. "Pray tell me what you want,
And I'll help you if I can?"
Then the blue eyes grew more eager,
And the breath came thick and fast;
And I saw within the chubby hands,
A folded paper grasped.

"Nurse told me," he said, "that the lightning
Came down on the wires some day;
And my mama has gone to heaven
And I'm lonely since she's gone away,
For Papa is very busy
And hasn't much time for me
So I thought I'd write her a letter
And I've brought it for you to see.

"I've printed it big so the angels
Could read out quick the name
And carry it straight to my Mamma
And tell her how it came,
And now won't you please take it,
And throw it up good and strong,
Against the wires in a funder shower,
And the angels will take it along."

Ah! What could I tell the darling?
For my eyes were filling fast;
I turned away to hide the tears,
But I cheerfully spoke at last:
"I'll do the best I can, my child,"
'Twas all that I could say;
"Thank you," he said, then scanned the sky;
"Do you think it will funder to-day?"

But the blue sky smiled in answer,
And the sun shone dazzling bright,
And his face, as he slowly turned away,
Lost some of its gladsome light;
"But nurse" he said, "if I stay so long,
Won't let me come any more;
So good-bye, I'll come and see you again
Right after a funder shower."

[Published 1891, Commercial and Railway Telegraphy - Author Unknown]

III.
FORMS - THE ENDLESS PAPER TRAIL

A station agent's duties were complex. The agent's job was not just sending/receiving messages or hooping orders to trains, it involved much more. The principle duties can be succinctly stated as: "The Agent is the local representative of the Freight, Passenger, Express, Communications and Industrial Departments. He is responsible to the Accounting Department for maintaining proper records, rendering statements and balance sheets. As the representative of the Treasurer, he is responsible for the collection of the Company's revenues."[20] The agent was the frontline employee to the public, representing all Officers and Departments of the railway, either directly or indirectly.

The telegraph agent and operator had many reports that constantly needed attention. Reports might be due hourly, daily, weekly, bimonthly, quarterly and yearly, and not just telegraphic reports. Reports would be completed between trains, telegram transmissions and customers. Canadian Pacific Communications noted that reports from station agents were not as numerous as other forms like express and waybills. Further complicating "Accounting and auditing not only of the Company's own telegraph receipts but also portions of revenue accruing to the Company from messages involving the Postal-Telegraph in the United States, and the Commercial Cable and Wireless Ltd. and other affiliated organizations throughout the world...."[21] Not in order or priority, Figure Three provides a list of the most common duties performed by the Agent.

FIGURE THREE:

SUMMATIVE DUTIES OF A STATION AGENT

A. GENERAL
- . good housekeeping and appearance of station grounds and interior
- . salesmanship
- . courtesy
- . handling complaints
- . station privacy [work/codes]
- . train bulletin boards
- . public notice posters
- . handling delays
- . switch alignments
- . placards on cars
- . station sanitation
- . police matters
- . station clock
- . destroy records/telegram copies
- . Agency Reports
- . requisition of supplies/work around station
- . writing business letters

B. FREIGHT DUTIES
- . Bills of Lading reports
- . checking inward and outward freight cars
- . freight reports and bills
- . loading cars properly
- . handling explosives and dangerous goods
- . freight tariffs, custom regulations
- . waybill bills
- . delivery of freight
- . car service operation
- . manifest fast freight
- . seals on cars
- . perishable notices
- . freight classes
- . loss and damage reports
- . damage prevention
- . perishable arrival

C. PASSENGER DUTIES
- . timetables
- . tariffs
- . fare classes
- . destinations, connections, stopovers
- . tickets, baggage, parcels, mail

D. EXPRESS DUTIES
- . classification and tariffs
- . accounting
- . handling

E. COMMUNICATIONS DEPARTMENT
- . railway service correspondence
- . delivery of telegrams
- . money orders
- . sending, receiving by telegraph, Telex, Telephone
- . accounting forms
- . commissions
- . tariffs
- . accounting
- . courtesy

- Compiled from CNR Handbook -

The agent was required to complete, as noted, as many as thirty different and related forms. Daily records were maintained to assist in the collection of hand written or typed reports.

CNT 6109
11-49

CANADIAN NATIONAL TELEGRAPHS
TELEGRAM RECEIVED BY TELEPHONE

RECORDERS NO. SIGN	CHECK	TIME FILED	SENDERS FULL NAME IF NOT SHOWN IN SIGNATURE
	40/36 NL DUEBILL		

TOLLS TO BE PAID BY OR NAME OF ACCOUNT TO BE CHARGED TO:	ADDRESS	SENDERS' PHONE NUMBER

JAN. 24/56

F BLACK

SUPT OF VETERANS INSURANCE DEPT OF VETERANS AFFAIRS OTTAWA ONT

RE VETERANS INSURANCE APPLICATION NO 41736 REFERENCE YOUR LETTER OF JANUARY TWENTY THIRD ETC.

ROBERT P. THOMPS

HOR 23

.50

not station to debit

CNT 6195
5-41

OTHER OFFICE COLLECTIONS

OFFICE FEB 8/56 19

MANAGER

ROBERT P THOMPS .50
TUMBLEWOOD
RR1.

THE ACCOUNT LISTED ABOVE HAS BEEN COLLECTED BY THIS OFFICE, AND YOU SHOULD NOW TAKE CREDIT ON YOUR NEXT BALANCE SHEET, ATTACHING THIS MEMORANDUM AS AUTHORITY.

THIS REFERS TO YOURS TELEGRAM JAN 24/56 TO C F BLACK, OTTAWA, ONT.

W. H. Roberts

THIS ADVICE TO BE FORWARDED IN DUPLICATE TO OFFICE REQUIRING CLEARANCE

Telegraph Forms used by the station agent. Note the detail required by the agent to account for every cent received or returned as change. Several forms used remained unchanged in style for years, for example, Form #4 was first prepared in August 1929 and was still in use in 1956.
(CNR Employee Handbook, 1956 - courtesy G.W. Horner)

CNT-6613
8-29

PRINTED IN CANADA

RECEIVED COLLECT MESSAGE RECORD

FOR PERIOD OF ________ ENDING ________ 19____ OFFICE ________

MESSAGE NUMBER	ADDRESSEE	PLACE FROM	MESS'GR NO.	CASH		CLERK'S SIGN	IN THIS COLUMN SHOW BILLING INFORMATION, SERVICED BACK, FRANK No., OR FORWARDED, ETC.
				TOLLS	TAX		
	TOTAL CASH ·	COLLECTED					

CNT-6617
10-47

CANADIAN NATIONAL TELEGRAPHS

CALCULATION OF COMMISSION

NOTE:
RAILWAY OFFICES HAVING ONE COMMISSION RATE MAY COMBINE THE "THIS LINE" DOMESTIC RECEIPTS AND "THIS LINE" DOMESTIC CHECKS, SHOWING ONE TOTAL FOR EACH.
INTERCHANGE: SEE TARIFF INSTRUCTION. COMMISSION ALLOWED ON 50% OF INTERCHANGE TOLLS.
ROUTINE FOR CALCULATING COMMISSION: SEE ACCOUNTING INSTRUCTIONS.

		COMMISSION	
EASTERN "THIS LINE" TOLLS			
TOTAL RECEIPTS	$ 24.30		
TOTAL CHECKS	$ 25.72		
COMMISSION @ 10	$ 50.02	5	00
WESTERN "THIS LINE" TOLLS			
TOTAL RECEIPTS	$		
TOTAL CHECKS	$		
COMMISSION @	$		
CONJOINT "THIS LINE" TOLLS			
TOTAL RECEIPTS	$		
TOTAL CHECKS	$		
COMMISSION @	$		
INTERCHANGE TOLLS (SEE NOTE)			
TOTAL RECEIPTS (50%)	$		
TOTAL CHECKS (50%)	$		
COMMISSION @	$		
MONEY TRANSFERS			
EASTERN RECEIPTS @ 10	$ 1.00		10
CHECKS @	$		
WESTERN RECEIPTS @	$		
CHECKS @	$		
CONJOINT RECEIPTS @	$		
CHECKS @	$		
REPORTS			
GENERAL PRESS @	$		
MARKET CABLES @	$		
FULL RATE SENT—No. OF WORDS	24 @ 1/2		12
FULL RATE REC'D.—No. OF WORDS	35 @ 1/2		17
DFD. RATE SENT—No. OF WORDS	48 @ 1/5		10
DFD. RATE REC'D.—No. OF WORDS	42 @ 1/5		08
	@		
	@		
TOTAL COMMISSION		5	57

DATE	No. OF MESSAGES (NOT INCLUDING FREE MESSAGES) CHECKED — SENT	CHECKED — REC'D.	RELAYED OR THRU
1	1	2	
2			
3	2	1	
4			
5		2	
6	4		
7			
8		2	
9	2		
10		3	
11			
12	1		
13		1	
14	3		
15			
16	4	4	
17			
18	2	2	
19			
20	1		
21		1	
22	3		
23			
24		1	
25	3		
26			
27	5	1	
28			
29			
30			
31			
TOTALS	31	20	
No. OF "RS" MESSAGES	3	3	

CERTIFIED CORRECT:

MONTH OF MAY 19 57
PERIOD OF MONTH

(SGD) JOHN DOE
AGENT

ABC, QUE.
OFFICE

PRINTED IN CANADA

CNT-6617
7-29

CANADIAN NATIONAL TELEGRAPHS

CALCULATION OF COMMISSION

EASTERN "THIS LINE" TOLLS
TOTAL RECEIPTS $______
TOTAL CHECKS $______
COMMISSION @___% $______

WESTERN "THIS LINE" TOLLS
TOTAL RECEIPTS $______
TOTAL CHECKS $______
COMMISSION @___% $______

CONJOINT "THIS LINE" TOLLS
TOTAL RECEIPTS $______
TOTAL CHECKS $______
COMMISSION @___% $______

INTERCHANGE "THIS LINE" TOLLS
TOTAL RECEIPTS $______
TOTAL CHECKS $______
COMMISSION @___% $______

______ "THIS LINE" TOLLS
TOTAL RECEIPTS $______
TOTAL CHECKS $______
COMMISSION @___% $______

REPORTS
GENERAL PRESS $______
MARKET______ $______
COMMISSION @___% $______

MONEY TRANSFERS
(AS PER FORM 6426)
EASTERN TOLLS $______
COMMISSION @___% $______
WESTERN TOLLS $______
COMMISSION @___% $______
CONJOINT TOLLS $______
COMMISSION @___% $______

CABLES AND MARCONIS
FULL RATE SENT - - WORDS @______
DEFERRED RATE SENT WORDS @______
FULL RATE REC'D - - WORDS @______
DEFERRED RATE REC'D WORDS @______

TOTAL COMMISSION

COMMISSION

DATE	NO. OF MESSAGES (NOT INCLUDING FREE MESSAGES)		
	CHECKED		RE-LAYED OR THRU.
	SENT	REC'D	
1			
2			
3			
4			
5			
6			
7			
8			
9			
10			
11			
12			
13			
14			
15			
16			
17			
18			
19			
20			
21			
22			
23			
24			
25			
26			
27			
28			
29			
30			
31			
TOTALS			

NO. OF "RS" MESSAGES

CERTIFIED CORRECT

____________________ AGENT

____________________ OFFICE

PRINTED IN CANADA

CNT 6618
4-48

AGENT'S CASH BALANCE SHEET

FOR PERIOD________________19____ OFFICE ________________

BINDING SPACE

LINE NUMBERS	DEBITS DESCRIPTION	AMOUNT	THIS COLUMN FOR USE OF R.A.C.	LINE NUMBERS	CREDITS DESCRIPTION	AMOUNT	THIS COLUMN FOR USE OF R.A.C.
1	BALANCE BROUGHT FORWARD FROM LINE 43 OF PREVIOUS BALANCE SHEET			23	BALANCE DUE AGENT FROM LINE 21 OF PREVIOUS BALANCE SHEET		
2	CASH RECORD TOLLS FORM 6200			24			
3	RECEIVED COLLECT MESSAGE TOLLS FORM 6613			25	OTHER LINE TOLLS & SPECIAL DELIVERY - FORM 6614 LOCAL		
4	MONEY TRANSFER PRINCIPAL FORM 6200			26	" " CONJOINT		
5	MONEY TRANSFER CHARGES FORM 6200			27	" " CABLE		
6	MONEY TRANSFER TOLLS INCLUDING EXTRA WORD TOLLS FORM 6200			28	COMMISSION - FORM 6617		
7	MONEY TRANSFER EXCHANGE FORM 6200			29			
8	MESSENGER ERRAND & DISTRIBUTION SERVICE - FORM 6515			30			
9	TAX - FORM 6200			31	MISCELLANEOUS CREDITS (DETAILS ON REVERSE SIDE) AUTHORITY ATTACHED		
10	TAX - FORM 6613			32	REMITTANCES: NO. DATE		
11	TAX - MONEY TRANSFERS			33	NO. DATE		
12	MISCELLANEOUS DEBITS (DETAILS ON REVERSE SIDE)			34	NO. DATE		
13	CABLE REGISTRATIONS FORM 6224			35			
14	ACCOUNTING ERROR NOTICES (DETAILS ON REVERSE SIDE)			36			
15	COLLECTIONS ON CUSTOMERS' ACCOUNTS			37			
16	COLLECTIONS ON DUE BILLS (DETAILS ON REVERSE SIDE)			38			
17				39			
18				40			
19				41			
20				42			
21	BALANCE DUE AGENT			43	CASH BALANCE ON HAND		
22	TOTAL (MUST AGREE WITH LINE 44)			44	TOTAL (MUST AGREE WITH LINE 22)		

________________19____ SIGNATURE________________
MANAGER OR AGENT

CANADIAN NATIONAL TELEGRAPHS

Accounting Department #83

____

_______________________194

Manager or Agent,

____

Herewith please find Customers' Accounts for month

of ________________ Totalling $____________ $____________

Early Accounts Rendered $____________

Total $____________

Ledger Control and Military accounts to follow.

Supervisor
Revenue Accounting Centre

8

CNT 6384
10 - 40

PRINTED IN CANADA

TELEGRAPHS CANADIAN NATIONAL CABLES

CERTIFICATE RE DESTRUCTION OF MESSAGES

(TO BE SENT IN EACH MONTH WITH MONTHLY STATEMENT)

IN COMPLIANCE WITH DIRECTIONS IN RULE 136 IN TARIFF BOOK, I CERTIFY THAT ALL SENT PAID, SENT COLLECT, RECEIVED PAID, RECEIVED COLLECT, RELAYED AND SERVICE MESSAGES AT________________________

________________________OFFICE

UP TO________________HAVE BEEN DESTROYED BY FIRE.

OFFICE________________

DATE________________ SIGNATURE________________
MANAGER OR AGENT.

CNT 6201
3-48

RECORD OF MESSAGES HANDLED FOR COMMISSION PURPOSES

OFFICE A B C, QUE.

DATE MAY 19 57

RECEIPTS								CHECKS							
SENT PAID & RECEIVED COLLECT MESSAGES								SENT COLLECT & RECEIVED PAID MESSAGES							
MSG. CHK.	PLACE TO OR FROM	THIS LINE TELEGRAM TOLLS				CABLES WORDS		MSG. CHK.	PLACE TO OR FROM	THIS LINE TELEGRAM TOLLS				CABLES WORDS	
		EAST	WEST	CON'J	INTER.	FR	DFD			EAST	WEST	CON'J	INTER.	FR	DFD
11	TORONTO, ONT.	74						16	BUFFALO, N.Y.			110			
10	REGINA, SASK.		140					13	PARIS, FRANCE						13
14	NEW YORK, N.Y.			91				14	THESSALON, ONT.				115		
18	LONDON, ENG.						18	18	MONTREAL, QUE.	74					
14	DUBLIN, IRE.					14		23	BIRMINGHAM, ENG.					23	
								10	CALGARY, ALTA.		170				
		1830	320	91	189	24	48			2017	330	110	115	35	42

NOTE. THIS FORM IS FOR OFFICE USE ONLY AND IS NOT TO BE SENT TO R.A.C. EXCEPT WHEN INSTRUCTED TO SEND MESSAGES FOR AUDITING PURPOSES.

INSTRUCTIONS: OFFICES ON TELETYPE OPERATION WILL LIST EACH RECEIVED PAID MESSAGE SHOWING "DATE", "PLACE" AND "TOLLS" IN APPROPRIATE COLUMNS UNDER "CHECKS". LIST TOLLS ONLY OF SENT COLLECT MESSAGES UNDER "CHECKS". SENT PAID AND RECEIVED COLLECT MESSAGES SHOW TOLLS ONLY IN APPROPRIATE COLUMNS UNDER "RECEIPTS". MAKE SEPARATE RECAPITULATION BY DAYS OF THE TOTAL TOLLS UNDER HEADINGS "RECEIPTS" AND "CHECKS". AND TRANSFER THE FINAL TOTALS TO FORM 6617, "CALCULATION OF COMMISSION".

OFFICES NOT ON TELETYPE OPERATION WILL LIST TOLLS OF EACH MESSAGE UNDER THE RESPECTIVE COLUMNS OF "RECEIPTS" AND "CHECKS" AND PREPARE RECAPITULATION ON SEPARATE FORM 6201 THE SAME AS OFFICES ON TELETYPE OPERATION.

MOD'S, GENERAL PRESS AND MARKET REPORTS (IF ANY) USE SEPARATE SHEET FORM 6201 INSERTING HEADINGS "MOD'S" ETC. UNDER "RECEIPTS" AND "CHECKS".

PRINTED IN CANADA

CNT 6619
8-41

STATEMENT OF UNCOLLECTED ACCOUNTS

A B C, QUE. OFFICE MONTH OF MAY 1957

NAME OF PATRON FROM WHOM DUE	ARREARS			CURRENT MONTH		HAS CREDIT BEEN CANCELLED? IF SO, DATE	WHY UNPAID AND WHEN PAYMENT EXPECTED
	MONTH	AMOUNT					
JOHN DOE & CO.	MAY	3	25				ACCOUNT SENT HEAD OFFICE FOR PAYMENT
HARRY SMITH,	"	12	25				OUT OF TOWN, WILL COLLECT ON RETURN
		--	--				
		15	50				
TOTAL						GRAND TOTAL	

PRINTED IN CANADA

CNT-6620
7-29

PRINTED IN CANADA

CUSTOMER'S LEDGER CONTROL SHEET

DATE MAY 19 57

OFFICE ABC, QUE.

"A" / "B" DATE	1 BALANCE FROM PREVIOUS MONTH AND BILLS RENDERED FOR CURRENT MONTH	2 COLLECTIONS AND OTHER CREDITS	3 DAILY BALANCE OUTSTANDING
A	15 50		
B	67 25		82 75
1		7 45	75 30
2		–	75 30
3		–	75 30
4		–	75 30
5		12 60	62 70
6		–	62 70
7		–	62 70
8		10 40	52 30
9		–	52 30
10		–	52 30
11		36 80	15 50
12			
13			
14			
15			
16			
17			
18			
19			
20			
21			
22			
23			
24			
25			
26			
27			
28			
29			
30			
31			
TOTAL	82 75	67 25	15 50

CNT-6620
7-29

PRINTED IN CANADA

Due Bill LEDGER CONTROL SHEET

DATE MAY 19 57

OFFICE ABC, QUE.

"A" / "B" DATE	1 BALANCE FROM PREVIOUS MONTH AND BILLS RENDERED FOR CURRENT MONTH	2 COLLECTIONS AND OTHER CREDITS	3 DAILY BALANCE OUTSTANDING
A	1 55		
B			1 55
1	70		2 25
2	85	1 55	1 55
3			
4			
5			
6			
7			
8			
9			
10			
11			
12			
13			
14			
15			
16			
17			
18			
19			
20			
21			
22			
23			
24			
25			
26			
27			
28			
29			
30			
31			
TOTAL	3 10	1 55	1 55

CNT 6191-D
11-37

DOIT TO / AU CANADIAN NATIONAL TELEGRAPHS DR.

NO. 1

MR. JOHN BROWN

ABC, QUE.

TOLLS ON TELEGRAM DATED / PRIX DE TÉLÉGRAMME ENVOYÉ MAY 15 19 57 $ .35 CTS.

TO — À
FROM — DE

RECEIVED PAYMENT
REÇU PAIEMENT

DATE ____ 19__ ____ FOR / POUR CANADIAN NATIONAL TELEGRAPHS

CNT-6105
8-49

FREE MESSAGE REPORT

A B C, QUE. (OFFICE) MAY (MONTH) 19 57

ALL SENT AND RECEIVED FREE AND PART FREE MESSAGES (INCLUDING THOSE UPON THE BUSINESS OF ANY RAILWAY COMPANY) ARE TO BE ENTERED UPON THIS REPORT, WHICH WITH THE MESSAGES MUST BE SENT TO THE REVENUE ACCOUNTING CENTER BY THE 5TH OF THE FOLLOWING MONTH.

WHEN NO FREE MESSAGES ARE SENT OR RECEIVED DURING THE MONTH, A "NIL" REPORT WILL NOT BE FORWARDED.

SENT FREE MESSAGES

ENCLOSE BOTH SENT AND RECEIVED MESSAGES WITH THIS REPORT

FROM WHOM	WHY SENT FREE (WHEN SENDER HOLDS FRANK, ENTER NUMBER OF SAME IN THIS COLUMN)	NO. OF MESSAGES
H. JONES	DH FRANK S 256	4
T. FRASER	DH EMP.	1

RECEIVE FREE MESSAGES

TOTAL NUMBER OF RECEIVED FREE MESSAGES 14

FOR INSTRUCTIONS REGARDING FREE MESSAGES, SEE RULES 18 TO 30 (INCLUSIVE), AND 140

CNT-6622
7-28
PRINTED IN CANADA

ACCOUNTING ERROR NOTICE

No. 1000

FROM REVENUE ACCOUNTING CENTRE AT MONTREAL, QUE.
To MANAGER OR AGENT AT ABC, QUE.

THE FOLLOWING ERRORS WERE FOUND IN CHECKING YOUR REPORTS FOR THE DATES SHOWN. THESE ERRORS MUST BE INVESTIGATED AND ACCOUNTED FOR OR EXPLAINED, AND THIS NOTICE RETURNED WITH YOUR NEXT BALANCE SHEET.

EXPLANATION MAY BE DETAILED BY MESSAGE REFERENCE NO. ON REVERSE SIDE.

DATE OF MESSAGE	REFERENCE NO.	NAME TO OR FROM	PLACE TO OR FROM	AMOUNT REPORTED	CORRECT AMOUNT	DIFFERENCE
MAR.7		J. SMITH	MONTREAL	35	55	20
			TOTALS			

CNT-6384
2-34

CERTIFICATE RE DESTRUCTION OF MESSAGES

(TO BE SENT IN EACH MONTH WITH MONTHLY STATEMENT)

IN COMPLIANCE WITH DIRECTIONS IN RULE 126 MAINLAND AND 140 NEWFOUNDLAND IN RULES AND REGULATIONS, I CERTIFY THAT ALL SENT PAID, SENT COLLECT, RECEIVED PAID, RECEIVED COLLECT, RELAYED AND SERVICE MESSAGES AT ABC, QUE. OFFICE

UP TO NOVEMBER 30, 1956, HAVE BEEN DESTROYED BY FIRE.

OFFICE ABC, QUE.

DATE JUNE 1, 1957 SIGNATURE (SGD) JOHN DOE
MANAGER OR AGENT

CNT-6200
10-50

CASH RECORD

OFFICE A B C, QUE. CLERK'S INITIALS ______ MONTH OF MAY 19 57

MESSAGES AND CABLES

No.	TOLLS		INITS.	No.	TOLLS		INITS.	No.	TOLLS		INITS.
1		65		1				1			
2	1	20		2				2			
3		70		3				3			
4		80		4				4			
5		64		5				5			
6		72		6				6			
7		35		7				7			
8		35		8				8			
9	5	41		9				9			
0				0				0			
1				1				1			
2				2				2			
3				3				3			
4				4				4			
5				5				5			
6				6				6			
7				7				7			
8				8				8			
9				9				9			
0				0				0			
1				1				1			
2				2				2			
3				3				3			
4				4				4			
5				5				5			
6				6				6			
7				7				7			
8				8				8			
9				9				9			
0				0				0			
							TOTAL				

SUMMARY

MESSAGES AND CABLES	5	41
MONEY ORDERS	51	60
RECEIVED COLLECTS	2	68
CUSTOMER'S ACCOUNTS	67	25
DUE BILLS	1	55
OTHER COLLECTIONS		
TOTAL CASH REMITTED	$ 128	49

MONEY ORDERS

CASH

DATE	PRINCIPAL		CHARGES		TOLLS		EXCHANGE		INITS
	50	00		60	1	00			
TOTAL									

OTHER COLLECTIONS

NAME	MONTH	AMOUNT	
CUSTOMERS' ACCOUNT			
1 JOHN DOE & CO.,	JAN.	37	45
5 HARRY SMITH,	"	8	05
5 R. ROBERTSON & CO.,	FEB.	4	55
8 A. JONES & CO.,	"	10	40
11 CAN. ACCEPT. CORPN.	"	6	80
		67	25
DUE BILLS.			
2 JOHN BROWN	FEB 27	1	55
TOTAL			

PRINTED IN CANADA

CNT 6200
5-41

CASH RECORD

OFFICE ______ CLERK'S INITIALS ______ 19 ___

MESSAGES AND CABLES					
No.	TOLLS	TAX	No.	TOLLS	TAX
1			1		
2			2		
3			3		
4			4		
5			5		
6			6		
7			7		
8			8		
9			9		
0			0		
1			1		
2			2		
3			3		
4			4		
5			5		
6			6		
7			7		
8			8		
9			9		
0			0		
1			1		
2			2		
3			3		
4			4		
5			5		
6			6		
7			7		
8			8		
9			9		
0			0		
1			1		
2			2		
3			3		
4			4		
5			5		
6			6		
7			7		
8			8		
9			9		
0			0		
1			1		
2			2		
3			3		
4			4		
5			5		
6			6		
7			7		
8			8		
9			9		
0			0		

MONEY ORDERS							
PRINCIPAL	CASH				CHARGE		
	CHARGES	TOLLS	EXCH.	TAX	CHARGES	TOLLS	TAX

OTHER COLLECTIONS		
NAME	MONTH	AMOUNT

SUMMARY

MESSAGES AND CABLES	
MONEY ORDERS	
OTHER COLLECTIONS	
TOTAL CASH REMITTED	

CNT 6434
12-42

STATEMENT OF MONEY ORDER APPLICATIONS FORWARDED TO AUDITOR, TORONTO

CAN. NAT. TELEGRAPHS PRINCIPAL		WESTERN UNION PRINCIPAL	
TOTAL			

THE ABOVE TOTAL AGREES WITH THE AMOUNT ACCOUNTED FOR ON REVENUE REPORT OF THIS DATE

PRINTED IN CANADA

OFFICE__________

REPORTING TO R.A.C.__________

DATE OF APPLICATIONS__________

NUMBERED FROM__________TO__________

CNT-6185
12-25

FREE MESSAGE REPORT

__________OFFICE MONTH OF__________19____

(WHEN NO FREE MESSAGES ARE SENT OR RECEIVED DURING THE MONTH, WRITE THE WORD "NIL" ACROSS THE FACE OF THIS REPORT, AND FORWARD SAME TO AUDITOR, TORONTO).

ALL SENT AND RECEIVED FREE AND PART FREE MESSAGES (INCLUDING THOSE UPON THE BUSINESS OF ANY RAILWAY COMPANY) ARE TO BE ENTERED UPON THIS REPORT, WHICH WITH THE MESSAGES, MUST BE SENT TO THE AUDITOR AT TORONTO, ONT., BY THE 5TH OF THE FOLLOWING MONTH.

ENCLOSE BOTH SENT AND RECEIVED MESSAGES WITH THIS REPORT

SENT FREE MESSAGES

FROM WHOM	WHY SENT FREE (WHEN SENDER HOLDS FRANK, ENTER NUMBER OF SAME IN THIS COLUM)	NO. OF MESSAGES

RECEIVED FREE MESSAGES

TOTAL NUMBER OF RECEIVED FREE MESSAGES__________

FOR INSTRUCTIONS REGARDING FREE MESSAGES, SEE RULES 18 TO 32 (INCLUSIVE), AND 90 AND 101

PRINTED IN CANADA

IV.
THE RAILWAY TELEGRAPH OPERATOR AND DISPATCHER

The telegraph operator's official title was "Railway Telegrapher" or "Operator." They did not have the Life of Riley depicted. It was a tough life. As noted, the duties and responsibilities involved required an individual of superior organizational skills, one with an even temperament, the ability to cooperate and work effectively with others, and a love for the job.

In the 1860s, the Grand Trunk Railway began formulating rules. Specifically, agents or operators had specific operational rules to follow. Operators who lived in the stations under House Fuel and Lights (HFL), had to be in their stations or offices whenever the company's interests demanded it. One hour was allowed for each meal, but rarely without interruption. An HFL agent was generally on duty twenty-four hours. The Grand Trunk often provided a house next to or as part of the actual station. Normal hours for the station required that it be open thirty minutes before the first train arrival and thirty minutes after the last expected one left. The agent was required for duty even if an extra train was due before or after closing time. Before an agent left the station, he had to notify the watchman, caretaker or dispatcher of their location if not an HFL station. The dispatcher had to give his consent for the leave. This was often a nuisance so a son or daughter replaced their father. In short, an operator's job could be subtitled, "Eternal Vigilance." Around junctions, terminals or busy mainline stations, there were shifts and sufficient coverage.

Some operator's were not as astute as others. Between 1844 and about 1865: "An operator who is not thoroughly competent to read by sound, should invariably allow the register-paper to run.... [and] In writing with offices where the register is used, a few dots shall always be made before commencing the transmission of a message in order to allow time for starting the paper."[22]

Operators between 1860 and 1900 (and as late as 1910), were expected to repair the telegraph line in an assigned section. On slow branchlines an agent could be gone for a week repairing the circuit while getting in some fishing and hunting. For instance,

> ... when the line got 'down' he [operator] hired a vehicle, and armed with repairing tools, followed the line until he found and remedied the interruption. He did not think it 'infra dignitatem' [beneath one's dignity] to climb poles, dig holes, cut branches, replace defective poles or pieces of line, or in short, do anything for the good of the business.[23]

After 1900, the railways began hiring lineman to do repair work on the circuits so the agent would be available in the office or station.

By 1900, new telegraphy rules appeared. The Grand Trunk Railway Rulebook of 1911 devoted a major section to the duties and responsibilities of an agent. Specific ones included:

> 1. They will use great care in adjusting, especially in damp weather, never opening the telegraph key until positive that the wire is not being used.
>
> 2. The contents of all telegrams must be held as confidential.
>
> 3. The greatest care must be taken in sending and receiving telegraph orders regarding the movement of trains.
>
> 4. Operators are required to keep a record of the time trains pass their stations, inform themselves in regard to approaching trains, and be prepared to furnish information promptly to such persons as should receive it.
>
> 5. Operators must read all messages carefully before sending to prevent delay in sending. No message will be transmitted from dictation or otherwise than from legible copy. When difficult words occur in messages, transmission must be slow and distinct.

6. Contention for circuit, and profane or obscene language upon the wires is forbidden.

7. The local battery must be kept clean and in good working order. The liquid should always cover the zinc, the loss by evaporation being replaced with clean water. The blue solution should always cover the copper, but never reach the zinc.

8. Operators will send to the station Lineman each month any old zinc or copper they may have on hand.

9. Telegraphers will not be required to:
- shovel snow or stencil cars.
- sift coal ashes.
- attend flower gardens and lawns.
- scrub, clean and/or disinfect stations and out-buildings.
- clean and disinfect cars.
- clean chimneys or clean stove pipes.
- clean furnaces or unload, cut or pile fuel.
- put on storm windows, storm doors, or porches, or remove same.
- perform other work not incidental to the usual work of a telegrapher.
- teach telegraphy nor admit students not members of the station staff to their offices, except on agreement.

- compiled from GTR 1911 Rulebook -

The telegraph was not significantly used for train dispatching until the 1860s. Initially train orders were communicated verbally then committed to memory by crews with specific meets or problem areas noted on paper. Trains moved by unreliable watches, preset timetables that spanned many different time zones, unspecified operational rules and luck. Signal communication was minimal, using only hands and arms, flags, and a ball on a pole around stations or terminals: when the ball was low on the pole, it was a stop; when it was high, the track was clear therefore the term ‘highball.’ Trains no doubt waited for hours in-the-hole for an oncoming train. Typically that superior train may not have kept to its assigned schedule due to mechanical failure, track damage, fuel depleted, and so on. It is certain that only those operating the train knew exactly where they were on the run - no one else could know. Not until the late 1890s were portable telegraph sets placed on all classes of trains.

When the telegraph began use as a railway communication tool, train crews distrusted it. “In spite of the fact that many engineers, accustomed to being loose with their trains, looked on the telegraph with disdain, this great forward step toward safety in train operations played a major role until superseded when block system of train control came into use about 1880.”[24] William Cooke in 1837 stated, “... with a telegraph the manager in his office at Paddington could live like a spider along the line ...”[25] keeping track of his train crews. Cooke and Wheatstone felt watching the employees was a positive aspect of the telegraph system, but the Great Western Railway (GWR) objected as it implied their company was slovenly.

As the telegraph gained notoriety, communication between trains and people, railway safety and train operations became vital. More trains were beginning to run in Canada and safe train operation became increasingly difficult. Along the Grand Trunk Railway between Toronto and Montreal in the late 1850s, trains began operating with telegraphic assistance. Movements were accomplished in a haphazard manner by a dispatcher through his operator(s) who telegraphed the required station. This manner of operation was known as the Single Order System. Succinctly, messages were transmitted to station operators who converted the code into a written train order. Next, the operator forwarded the orders to the train conductor whose movements were affected. Early train orders had no standard procedure to define a meeting location for two trains; an order might be omitted, or another fatal omission would surely occur.

FORM—(C.)

GRAND TRUNK RAILWAY SYSTEM

CAUTION CARD No. 9

Station D ; 9.15 A. M. April 1, 19 11

Conductor and Engineer No. 26 on Eastbound Track

No. Further orders for your train.

Proceed with caution expecting to find track obstructed.

John Jones Signalman.

This card does not interfere with or countermand any orders you may have received. It is authority to pass the signal and proceed prepared to stop within the range of vision.

Engineer and Conductor must each have a copy and see that it is properly filled out and signed and that their train is correctly designated.

(Print on yellow paper.)

FORM—(D.)

GRAND TRUNK RAILWAY SYSTEM

CLEARANCE CARD.

(C) 9.05 a.m. June 3, 19 11

Train No. 12

Signal is inoperative. Proceed.

John Jones.

Signalman.

This does not affect any train order you have received.

(Print on white paper.)

NOTE.—Forms A and C may be combined in one form.

Telegraphic Forms from the Grand Trunk Railway Rulebook, 1911. The Agent would either write or type in the blank area. All telegraphic messages received by the agent from the dispatcher had to be verified precisely. These forms were either handed to or hooped up to the train crew depending on the order board displayed.
(R.G Burnet)

Train Order Blank for 31 Order.

FORM 31 | FORM 31

GRAND TRUNK RAILWAY SYSTEM

TRAIN ORDER No. 10

March 21, 19 11.

To .... | At ....

x ....(Initials)....... Opr., ... 1.45 A. .... M.

Conductor and Engineer must each have a copy of this Order.

Repeated at 2.20 A. *M.*

Conductor.	Train.	Made.	Time.	Operator.
Jones	45	Complete.	2.20 a.m	Black.

Specifications for Train Order Form and Books for Operators for "31" Orders.

Train Order Blank for 19 Order.

FORM 19 | FORM 19

GRAND TRUNK RAILWAY SYSTEM

TRAIN ORDER No. 10

March 11, 19 11.

To .... | At ....

x ....(Initials)....... Opr.; .... 1.45 A. .... M.

Conductor and Engineer must each have a copy of this Order.

Made Complete *time.* 1.16 P. *M.* Black *Opr*

Specifications for Train Order Form and Books for Operators for "19" Orders.

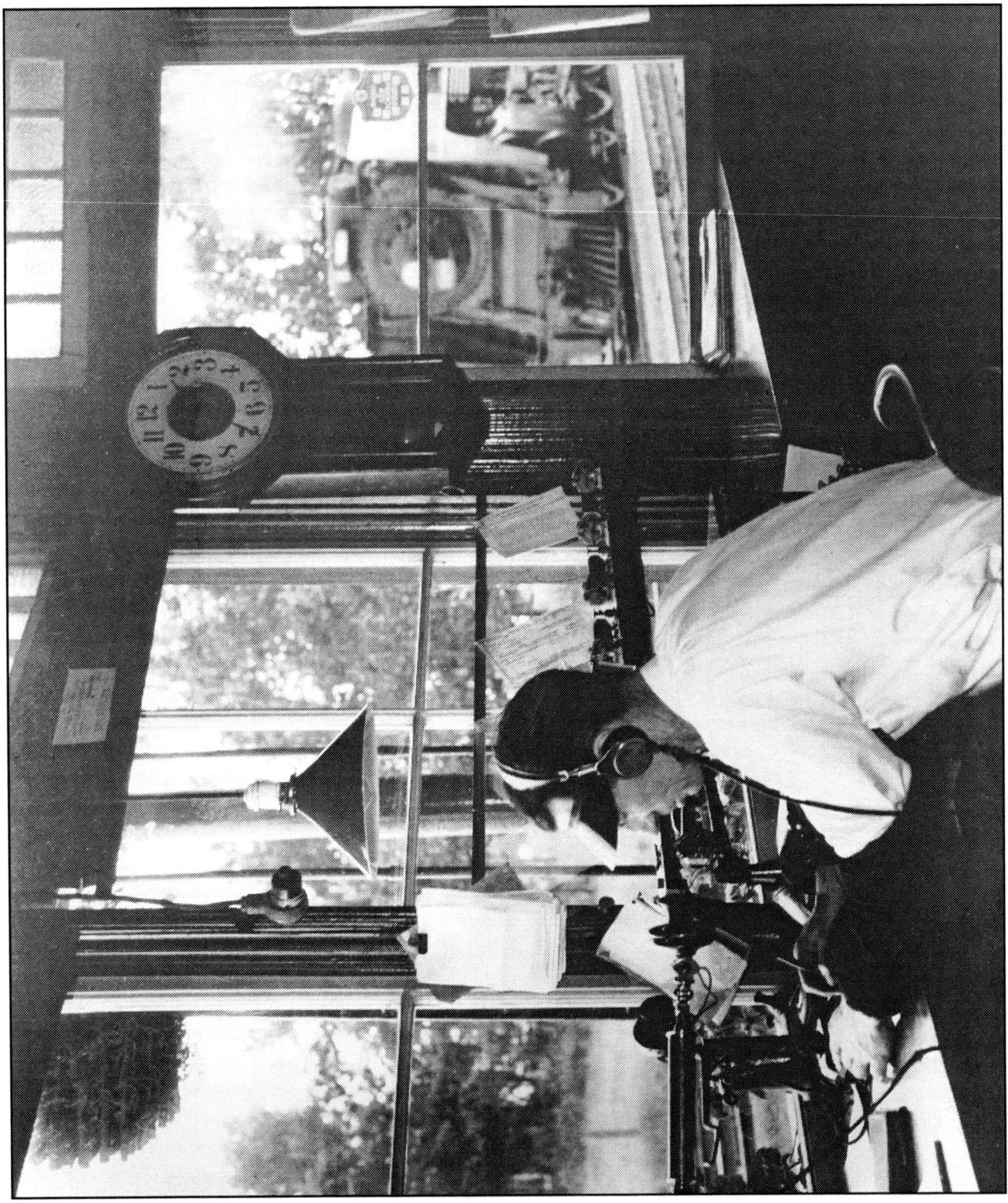

Operator Herman Snider at Gananoque Station around 1942. Note the intrusion and more prominent position of the telephone over the telegraph.
(CNR Public Affairs, #14346)

A family photograph of Spencer McCalpin (author's uncle) in 1930 at Danforth (York) station. Note the use of the "bug", preferred by many operators over the Bunnell telegraph key.
(R.G. Burnet)

The Double Train Order System developed after 1880. As the number of trains increased risks multiplied. This telegraphic operational procedure involved sending the same order simultaneously to all stations and trains whose movements were being effected. Therefore, two or more operators would copy the same train order. Next, the orders were repeated back, verbatim, to the dispatcher who verified the contents by underlining the repetition in his register book. If an error was detected, the procedure was repeated and the copy destroyed. This was done well ahead of every train's arrival in case of omission, misunderstanding or error. After the orders had been accurately verified, it was up to the operator to ensure the train crew received them. In terminal centres, the train crew met and read the orders to each other to be sure all understood; when hooped up, they were read on-the-fly.

The next step in train dispatching by telegraph came with the formalization of telegraphic train order forms. The Association of American Railroads was the first to create conventional telegraphic forms. However, individual needs of different companies were allowed. The general content of the forms remained the same making it easier for companies to interchange traffic and when employees moved to different roads. Their primary objective was to obtain absolute train movement safety in a succinct clear form. As noted earlier, Minot was the "Father of the Telegraphic Train Order."

Throughout their history, operators and dispatchers had to be fully competent with the rules. An expert knowledge of timetables, signals and the General Rules of the company was demanded. There were train orders for single track, giving rights over other given trains, rights to opposing trains, extras, work extras, holding orders, annulling parts of orders, superseding a full or part of an order. Double track had similar rules only different actions required.

The telegraph provided unique services to rural areas and small towns. In emergencies, station operators could be relied on for help. They also learned of eventful news, gossiped (though unofficial) and sent messages about important people, dangerous cargo, or the arrival of the circus train. Because a telegraph message was faster than the mail, the railway station took an unofficial role as an information centre. It, too, was a place for rural gatherings, to learn of news or wait for a train. Accordingly, an operator's position has been described as a job responsibility that stood between 'no trouble and bad trouble.'

There were heroes on the railway. Many have gone unnoticed and unrecorded, and likely they would want it that way. Old time railway men were crusty, yet extremely generous, kind and modest. One such event recorded was the 6 December 1917, 8:45 a.m., disaster in Halifax Harbour. The Belgian relief ship "Imo" was about to ram the French munitions ship "Mont Blanc." William Lovett, the chief clerk at the Richmond Yard saw the immediate danger and began warning people to take cover. Vincent P. Coleman, the train dispatcher, realizing a passenger train was due, returned to the station to send: "Hold up the train. Ammunition ship afire in harbour and will explode. Guess this will be my last message. Goodbye boys."[26] Copies of the Halifax newspaper, the Herald, praised the unselfish heroism of Coleman. (A recent television commercial has also immortalized this heroically tragic event.)

There are also many poems and songs about the aspects of the job. For example:

THE TELEGRAPHER'S LAMENT

It's not my job to run the train ... The
whistle I can't blow ... It's not for me to
say how far ... The train's allowed to go ...
It's not my place to shoot off steam ... Nor
even clang the bell ... But let the damn
thing jump the track ... And just see who
catches hell. [Anon.]

THE TRAIN DISPATCHER

You may talk about the engineer,
    So brave, so true and strong,
And say he is the only man
    That gets the train along.
Oh yes, there's the sweet conductor
    Who puts on such a dike
And is smiled on by all the ladies
    All along the pike;
But, my friends, let me tell you
    It's the working of the brain
Of the man who slings the lightning-
    He's the man who runs the train.

[Anon.]

CNR Dispatcher in Montreal taken about 1950. Note the solid telephone transmitter arm and especially the legless dual telegraph keys preferred by dispatchers, and the selector box with the buttons to call individual stations. The train sheets, train order book and blotter are timeless relics now.

(CNR Public Affairs, #32489)

V.
THE CHIEF WIRE OPERATOR

The Chief Wire Operator, or Circuit Manager, coordinated commercial operations, railway traffic, and dealt with specific troubleshooting problems. Typically,

> The desk of the Chief Operator is placed on a raised platform beside the door, so that he can observe the whole operating room. Beside him is the cabinet containing the battery switches & motor generators; next to this is the main switch case, which is extended up to the ceiling, forming a small room behind it, into which all the wiring of the office is concentrated. The outside wires are brought through a window in 19 conductor cables; each conductor ends on a porcelain fuse block, to protect against any abnormal current which may be brought in the office by outside contact with electric light or power wires.[27]

In the main office or terminal, the operating room was a constant busy centre. It was organized chaos, filled with equipment and little room to move around. The description of the CPR main office at the corner of McDermott and Main Street, Winnipeg, in 1899, provides a lively and typical set up:

> In the operating room the floor is hard maple, polished & oiled. There are 9 quartette operating tables, providing space for 36 operators, equipped with 7 sets of duplex, 1 quadruplex, 2 half sets of single repeaters & 16 sets of single-line instruments. The tables are of white wood with cherry tops, & are cut so that a portion in front of each operator can be removed & placed at a lower level to permit of the use of a type-writer. All the operators in this office are expert type-writers, & the bulk of the business is received on that machine. Formerly the sound of different instruments on each table was separated by means of glass partitions, & the instruments raised on small shelves, the top of the table being also cut into 4 parts so as to prevent it acting as a sounding board. This has all been done away with in the new office, the reading instruments (sounders) being placed in small boxes (resonators) elevated on an iron standard; these boxes have a curved top, & are open at the side next to the operator so that the sound is thrown out to him only, & does not become confused with that of other instruments.[28]

Located behind the operating room was the Battery Room. The new battery room in the CPR Winnipeg office was nine by seventeen feet. The previous room was in the basement and filled to capacity with 2,000 gravity style batteries. Used now, was "... 301 small cells of chlorine accumulator storage battery, divided into 7 equal banks. There are also 8 cells of a larger size working the sounders, 2 cells being in use while another bank of 2 are being charged, the remaining 4 cells are used for the purpose of extending the duplex circuits to the station about 1 mile away."[29] Also located in this room were two motor generators. One generator was used to drop the incoming 500 volts to 120 volts used to charge the main battery banks. The second generator lowered 500 volts to seven volts and was used to charge the local batteries needed for the sounder, relays and smaller equipment.

Figure Four highlights the people employed to get the message through:

FIGURE FOUR:
CLASSIFICATION OF EMPLOYEES IN THE COMMUNICATIONS DEPARTMENT - CANADIAN PACIFIC RAILWAY - 1937[30]

Assistant Manager
Assistant to the Genl. Mngr.
Acting Chief Engineer
Agent
Assistant Agent
All Night Chief
Acting Chief
Accountant
Assistant Accountant
Asst. Wire & Traffic Chief
Branch Agent
Broadcast Attendant
Broadcast Supervisor
Bookkeeper
Chief of Tariff Bureau
Chief Clerk
Clerk
Chief Operator
Assistant Operator
Night Operator
Check Clerk
Number Check Clerk
Clerk (Junior)
Cashier
Climber
Chief Collector
Collector
Chief Delivery Clerk
Chief Night Delivery Clerk
Carrier Attendant
Charwoman
Cableman
Cable Tester
Commission Agent
Caretaker
Draughtsman
Evening Chief
Engineer
Groundsman
Inspector
Installer
Linesman
Mechanic
Assistant Mechanic
Auto. Mechanic
Mux Supervisor
Asst. Mux Supervisor
Messenger
Mux Typist
Night Watchman
Night Chief
Night Chief Clerk
Office Boy
Operator
Phone Supervisor
Asst. Phone Supervisor
Night Phone Supervisor
Phone Clerk
Printer Traffic Chief
Chief Printer Supervisor
Repeater Attendant
Radio Supervisor
Repeater Chief
Rate Clerk
R. S. Night Traffic Chief
R. S. Day Traffic Chief
Radio Engineer
Route Clerk
Supervisor
Stenographer
Special Radio Representative
Supt. of Traffic
Superintendent
Service Supervisor
Test Agent
Typist
Traffic Chief
A.N. Railway Chief
Chief Ticker Supervisor
Ticker Supervisor
Ticker Attendant
Telephone Engineer
Telephone Maintainer
Timekeeper
Typist Mechanic
Traffic Solicitor
Asst. Traffic Solicitor
Transmission Engineer
Wire Chief
Night Wire Chief

Toronto 'C' Office, 347 Bay Street, Canadian National Telegraphs ca1940 - twelveth floor, Morse Department. Note the modernization that has occurred - the Underwood typewriters at each work station; the sounder on a three postion arm; the few 'old-timer's' wearing the green eye shades; only one woman operator in the second row; the message trough in the foreground replacing the brass pickup-arms; the extensive pneumatic air tubes for delivering telegrams and messages to the many floors in the CN Telegrams building and to other railway locations. In the background is the raised Wire Chief's desk.

(National Archives of Canada, PA 189970)

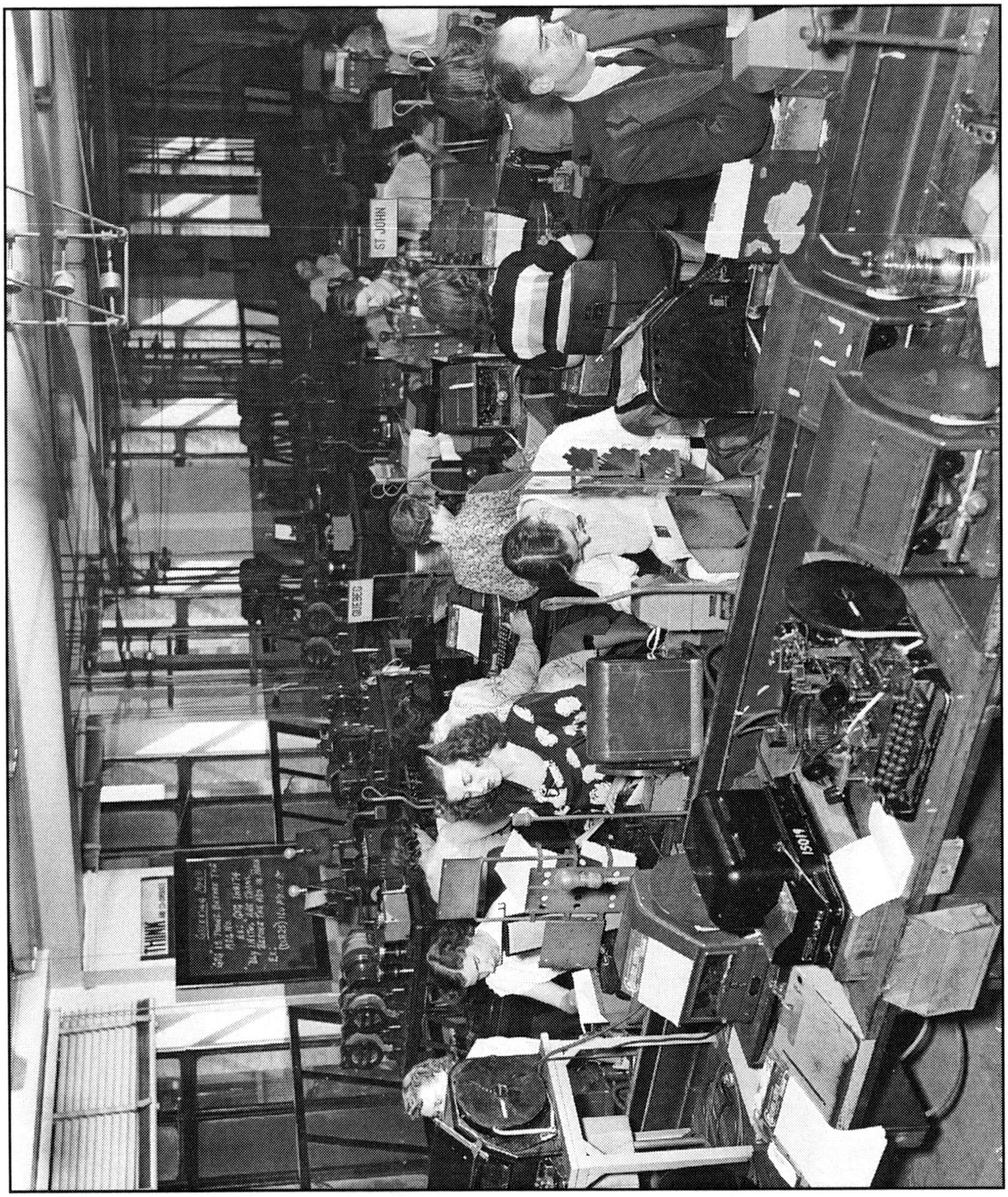

A view of how the Wire Chief might see the Automatic or Teletype and MUX Department, 10th Floor in the National Building at 347 Bay Street, Toronto from his raised platform. Note the various style of equipment used - teletypes, typewriters, message hooks, sounders. Against the far wall are a series of pneumatic tubes used to send messages around Union Station, the CNR Building and other CN office locations in the city. This is the main operating room in Toronto about 1940. Floors one to nine were for officals, Revenue, Accounting and other offices. Floor Ten was the MUX or Automatic department as shown; floor eleven, the telephone recording floor, and the twelveth floor was the Morse Department.

(CNR Public Affairs, #X32089)

CONCLUSION

In 1929 J.J. Thornton wrote in the August 1929 Issue of Canadian National Magazine:

> Telegraph history is written in the demand the public makes for service. It is a demand that brooks no obstacles.

Later, in 1937, Canadian Pacific published a book of CPR Facts and Figures. It was written:

> Time continues to indicate that the transmission of the written word by telegraph is indispensable to trade, industry, and social life. Its reliability, coupled with its accuracy and speed, undoubtedly will continue to retain for it a popularity over all other forms of written communication.... Looking toward the future, there seems to be no more reason to believe that wires and cables will be supplanted by other means of communication than the motor cars and trucks will supersede the railway train.

The people who miss the telegraph most are the operators, dispatchers and those employed to use the equipment. No one likes to give up or give in. To retain the past, poems, songs, magazine articles, books and special groups and clubs help to maintain the experiences of the past. But they will never capture the true working aura.

Generation upon generation have been telegraph operators. It was often a family tradition and a job with status. Signing your name and occupation as a Railway Telegrapher to something as simple as a Birth Certificate, caused eyebrows to rise, smiles to form, and scowls to lighten on the sourest of people. A telegrapher's job commanded respect.

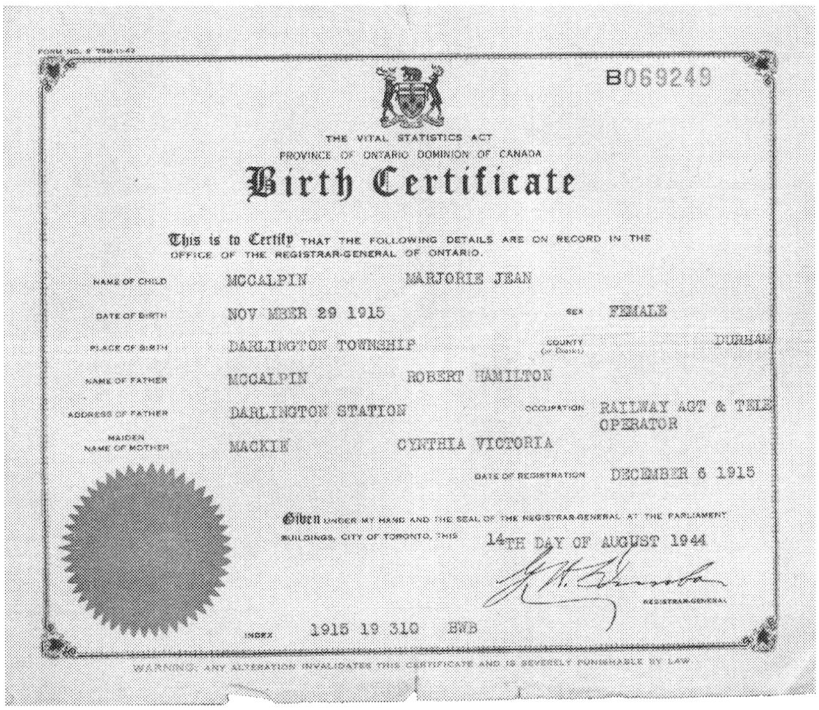

B069249

THE VITAL STATISTICS ACT
PROVINCE OF ONTARIO DOMINION OF CANADA

Birth Certificate

This is to Certify THAT THE FOLLOWING DETAILS ARE ON RECORD IN THE OFFICE OF THE REGISTRAR-GENERAL OF ONTARIO.

NAME OF CHILD: MCCALPIN MARJORIE JEAN
DATE OF BIRTH: NOV MBER 29 1915 SEX: FEMALE
PLACE OF BIRTH: DARLINGTON TOWNSHIP COUNTY (or District): DURHAM
NAME OF FATHER: MCCALPIN ROBERT HAMILTON
ADDRESS OF FATHER: DARLINGTON STATION OCCUPATION: RAILWAY AGT & TELE OPERATOR
MAIDEN NAME OF MOTHER: MACKIE CYNTHIA VICTORIA
DATE OF REGISTRATION: DECEMBER 6 1915

Given UNDER MY HAND AND THE SEAL OF THE REGISTRAR-GENERAL AT THE PARLIAMENT BUILDINGS, CITY OF TORONTO, THIS 14TH DAY OF AUGUST 1944

REGISTRAR-GENERAL

INDEX 1915 19 310 EWB

WARNING: ANY ALTERATION INVALIDATES THIS CERTIFICATE AND IS SEVERELY PUNISHABLE BY LAW

This certificate was reissued 14 August 1944 to my mother. The original was burned when the GTR Darlington Station where she was born (as all family members were between 1845 and 1915), burned down suddenly in late 1919. The station was not replaced and area passengers went to CNR Bowmanville. My grandfather then became Agent at Port Union until his death in 1931. Note my grandfather's occupation at Darlington: 'Railway Agt [Agent] and Tele[graph/phone] Operator.'

(R.G. Burnet)

No one in the railway telegraph industry really saw the demise of the telegraph. It was accepted, like the steam locomotive, that it would "just always be there 'cause it always had." Then one day it was replaced. No one would or could admit to it. It just happened.

Now with its passing, another segment of our railway heritage has virtually disappeared. The railways are an essential, viable, dependable, customer oriented business. Where improvements can be made for the customer, it will be done. The survival of business and industry depends upon it. There are little romantics in progress. All that can be done is for groups to preserve the past, not to live it, but to sanctify its historical significance. It helps everyone to know from where we came and to challenge the future with better services and products.

It was written in 1898 by J.M. Mackenzie in The Railway Station-A Social History, which describes the telegraph and its tremendous significance to industry and our cultural past best:

> The telegraph, though it occasionally blunders, is the swiftest and most zealous of all railway servants. It gives word of warning to the signal man, a hint of danger to the driver, a peremptory instruction to the station-master in emergency; it speaks even to the shunter amid the maze of waggons, and to the platelayer busy with his gang on the creosoted sleepers and rusty rails in the lonely cutting. It marshals goods trains, stops expresses, orders special trains, helps every official on the line....

Finally, Jim Hoell wrote the following poem several years ago titled "That Old Bug of Mine." It appeared in Railroad Magazine during the late 1930s. It shows the fondness that all can share and understand about this superlative technology and the people who performed it:

Well, old bug, this is the end of the road,
When no longer we transmit the old Morse Code;
We've flashed the final "Thirty" on this day,
And for the last time said "73-RJ."
You are a pile of junk and you look like hell,
Discoloured with age and your points worn well;
Your cord is frayed and your legs won't hold,
And like me, by gosh, you are growing old.

Year after year you fought snow, rain and fog
When the damn wires were heavy as a log;
Wind storms and flood held no terrors for you;
You tried your damnest to put your stuff through.
Over the high Sierras and Rockies, too,
Across deserts and plains your signals flew;
Through jungles and past many tow'ring peaks
You flashed a language which only Morse speaks.

The time has come when you and I must part
And I feel regret deep in my heart;
Pension time is here; I must step aside
And no more will my fingers be your guide.
Now you will lay on the shelf in the dust,
With a million memories as you rust;
But I know when Gabriel blows his horn
You will "OS" me by that judgment morn.

The telegraph lines and poles are chiefly telephone circuits now. And, even they are losing favour to micro- and macro-technology. Nick Morant's photograph of The Canadian clearly identifies the communication lines that first bounded Canada from east to west coasts. Lines which carried news, stock market quotations, world events, births, deaths, sport scores, train movements and railway business. Disappearing quickly is the unique sighing and singing harp string sounds of copper wires in a breeze; and the rhythmical staccato of the sounder with the gentle tapping of the key has almost faded completely. The importance and significance of the telegraph to Canada as a nation and to its railways can not be ignored.

(Courtesy Canadian Pacific Limited, Image No. M8252, Photographer N. Morant)

CANADIAN TELEGRAPH CHRONOLOGY
(Non-Canadian Event indicated by an *)

1846	Jul 24	First electric telegraph demonstration at Toronto City Hall
	Oct 22	Canada's FIRST telegraph company, the Toronto, Hamilton, Niagara & St.Catharines Telegraph Co. incorporates
	Dec 19	First political message from Toronto to Hamilton Mayor
		. The Toronto, Hamilton, Niagara & St. Catharines Telegraph Company opens for business from 22 Front Street, Toronto
	Dec 26	Montreal and Toronto Magnetic Telegraph Company organizes
		*. Royal E. House develops printing telegraph
1847		Montreal Telegraph begins telegraph construction west to Toronto and east to Quebec
		. Andrew Shaw appointed first president of Montreal Telegraph
	Jun 1	Telegram rate set at 25 cents for ten words and one cent more each additional word
	Jul 28	Montreal and Toronto Magnetic Telegraph Company obtains charter / name changed to Montreal Telegraph Company
		. British North American Telegraph Company incorporated
	Aug 3	Montreal Telegraph completes circuit Montreal to Toronto
	Oct 2	Montreal Telegraph completes circuit Montreal to Quebec
	Oct 4	New Brunswick Electric Telegraph Company incorporates
1848		First suggestion of an All-Red Route for the British Empire
1849	Feb 21	express horseback rider used between Halifax and Digby with news dispatches
	Nov 8?	last ride of the express horseback rider
	Nov 9	Circuit from Halifax to Portland and New York opens
1850		* F.C. Bakewell develops copying telegraph
1851		F.N. Gisbourne completes first telegraphic connection between St. John's and Conception Bay, Newfoundland
		Hugh Allan nominated and elected new president of the Montreal Telegraph
	Apr 8	* Western Union Telegraph Company opens agency
	Apr 30	Telegraph line established between St. John and St. Andrew's, New Brunswick
	Nov?	British North American Telegraph Co. service discontinued
	Nov 9	Nova Scotia Electric Telegraph incorporates
1852		Grand Trunk Railway of Canada builds tracks west from Montreal to Toronto - Montreal Telegraph lands the railway telegraph construction contract
	Jul 29	Montreal Telegraph takeover of the THN & St.C confirmed
	Nov 10	Act to provide one General Law for the Incorporation of all Electric Telegraph Companies in Canada
	Nov 22	First North American submarine cable operates between Prince Edward Island and New Brunswick
	Dec 14	Canada Grand Trunk Telegraph Company first company to incorporate under new General Law for Telegraph Companies (no affiliation with Grand Trunk Railway)
1853		Grand Trunk Railway obtains power to construct telegraph and small use of telegraph for train dispatching
	Nov 18	Montreal Telegraph obtains agreement with Grand Trunk Railway to build telegraph line beside their tracks
		* Sir William Werner Siemens develops effective insulation
		* Wilhem Gintl first proposes Duplex circuit transmission

1854	May 6	Cryrus Field of American Telegraph Company (USA) establishes New York, Newfoundland and London Telegraph Company in anticipation of trans-Atlantic submarine telegraph cable
1855		Montreal Telegraph acquires Canada Grand Trunk Telegraph
	Jun 2	Northern Railway (OS & H) reaches Collingwood
		* David Edward Hughes develops the stock ticker
1856	Nov 10	Telegraph and submarine cable service between St. John Newfoundland, Halifax Nova Scotia and New York, New York opens under the New York, Newfoundland and London Telegraph Company
1857	Aug 7-11	* First attempt to lay Atlantic Ocean submarine cable from Ireland to Trinity Bay, Newfoundland - fails
1858		North-West Transportation, Navigation & Railway Company given power to establish telegraph lines but not done - fails again in 1862 then abandoned
	Jul 7	Between July 7 and August 5 the second Trans-Atlantic cable is attempted but fails due to insulation in October
1860		Watkin suggests Grand Trunk Railway build telegraph and railway across northern Ontario to British Columbia / American Telegraph Company leases Nova Scotia Electric Telegraph Company
1862		Watkin suggests organizing an all-Canadian telegraph company to build from Ontario to British Columbia
1863	Jun 15	Agreement reached to purchase Rupert's Land
	Jul 6	Watkin heads west to initiate telegraph and railway plans
1864		Dr John Rae begins western survey for telegraph-railway route in summer
1865	Apr 18	* Collins Overland Telegraph Company connects New Westminister BC to mid-California
	Oct	Sandford Fleming takes over construction of Truro to Pictou Railway establishing telegraph ahead of railway along the proposed route
1866	May 31	Fenians attack Fort Erie, Ontario cutting telegraph circuit of Lake Huron and Buffalo Railroad
	Jul 27	Third attempt at a Trans-Atlantic cable is successful
	Aug	* Collins Overland Telegraph Company has lines to Yale and Hazelton
1867	Jul 1	George Blake, Manager of Hamilton Montreal Telegraph Office, first to experiment with electrical lighting in Canada; uses carbon pencils as a display on Dominion Day
1868		Dominion Telegraph Company begins telegram rate war with the Montreal Telegraph Company
1869	Sep 9	Montreal Telegraph obtains exclusive operating rights of telegraph system and construction for Grand Trunk Railway
1870		Western Union sells all Collins Overland Company lines to the Dominion Government
	May 14	Montreal Telegraph secures rights to build from Riviere-du-Loup to Halifax providing competition to Western Union
	Sep 22	Montreal Telegraph, Crown and Intercolonial Railway obtain exclusive telegraph agreement along ICR tracks

1871	Jan 11	Dominion Telegraph Company forms / obtains charter on 14 April 1871
	Jun 1	Montreal Telegraph announces free of debt
	Nov 20	Prairies have their first telegraph system built by the Northwestern Telegraph Company
1872		Western Union purchases all telegraph circuits in Nova Scotia
		* Jean Maurice Emile Baudot invents multiplex printing telegraph system
1874	Feb 19	J.D. Edgar instructed to obtain agreement with British Columbia for road and telegraph communication
	Sep 17	basic agreement for telegraph and railway to British Columbia accepted Oct 17 Contract One awarded by Dominion Government for Western Telegraph
	Oct 30	Contract Two awarded by Dominion Government for Western Telegraph
	Nov 10	Contract Three awarded by Dominion Government for Western Telegraph
1875	Feb 9	Contract Four awarded by Dominion Government for Western Telegraph
	Feb	Dominion Government contract for telegraph line from Prince Arthur's Landing to Selkirk
1878		Telegram rate war between Montreal and Dominion Telegraph Companies - now 20 cents for ten words - on the Prairies a ten word message ranges between $2.00 and $3.00 Spring - CPR telegraph Selkirk to Leduc
	Jul 25	Last Montreal Telegraph circuit and stations built opens between Matane and Fox River, Quebec
	Aug 28	Telegraph built from Prince Arthur's Landing to Winnipeg
1879		Sandford Fleming talks about and All-Red-Route by submarine cable and telegraph line from Great Britain, through Canada and to Australia and New Zealand. CPR connects Hays Lake to Leduc
	Jan 24	Edmonton has Dominion government hookup from Winnipeg to Battleford
	Jul 1	Dominion Telegraph leased to American Union Telegraph Co.
1880	May 7	Great North Western Telegraph Company incorporates and makes connections with Ontario and Quebec from Winnipeg head office
1881		CPR obtains all Dominion Government telegraph lines in British Columbia amounting to 181 miles
	Feb 15	Canadian Pacific Railway receives its railway Charter
	Jun 10	Erastus Wiman nominated and elected president of the Great North Western Telegraph
	Jul 1	Great North Western Telegraph obtains Montreal Telegraph
	Aug 17	Montreal Telegraph, Western Union now run by the GNWTCo . telegram rate restored to 25 cents for ten words
1882		CPR has 895 miles of telegraph lines operating . CPR Winnipeg to Port Arthur first CPR telegraph wire from east and intermediate points started / CPR connects Humboldt to Edmonton
1883		CPR begins commercial telegram service
	Feb	CPR has 897 miles of telegraph lines on Prairies
	Jul 1	CPR telegraph line from Winnipeg to Prince Arthur's Landing
1884		CPR telegraph into Laggan, Battleford and Edmonton. Riel Rebellion nets CPR $145,000 in telegraphic profits
1885		CPTelegraphs begins expanded commercial service between Lake Superior and the Rockies / breaks Western Union monopoly / lines extending from Port Arthur to Nipigon, Ontario
	Jan 24	CPR telegraph completed from Atlantic to Pacific

1886 CPR establishes four telegraphic divisions (1) 'Pacific' is British Columbia (2) 'Western' is Port Arthur to BC (3) 'Ontario' is Smiths Falls to Windsor and (4) 'Eastern' Port Arthur to Atlantic Cable in Nova Scotia

Jan 1 CPTelegraphs makes connections with east-west USA coast as well as opening head office in Montreal / leases two wires from Western Union and two other wires between Toronto and Niagara where the Canada Mutual Telegraph Company once operated

Jun 28 Official opening of Canadian Pacific Railway coast-to-coast

Sep 1 First segment of the All-Red-Route now in operation without breaks in circuit or need to use other companies

Dec 20 first segment of the all-Red Route officially opened from New Westminster, BC to Old Westminster, England

1887 CPTelegraphs constructs line from Walsh, Alberta to USA border and connects with Rocky Mountain Telegraph Company

Apr *Circuit opens between New Westminister B.C., and southern California on Postal Telegraph

* Oliver Heaviside shows how to minimize induction

1888 CPR makes connections with Frederiction, New Brunswick through Maine

1889 Feb 17 Line extended from Frederiction to Halifax and New Glasgow Nova Scotia after obtaining I.C.R. permission and direct link with Vancouver

Dec 18 CPR telegraph connects Canso, Nova Scotia with Great Britain by cable submarine cable under the Atlantic Ocean CPR moves Pacific Division Headquarters to Vancouver from New Westminster

1890 Railways standardize use of American Morse code

1894 CPTelegraphs connects with Associated Press to distribute news along the entire system

1896 Storage Batteries replaced with electric generators in Ottawa

Sep 27 CPR telegraph operators and dispatchers went on strike for shorter hours and higher wages - lasted to 7 October 1896

1898 CPR first Canadian company to use copper wire, replacing galvanized iron

1899 CPR Eastern Telegraph Division split in two at Montreal Western Division revised with Central Division

1900 * Horace G. Martin develops the "bug" or Vibroflex key

1901 CPT buries first underground cables in Montreal to relieve telegraph pole congestion in streets

1902 May 15 Canadian Northern Railway establishes its own telegraph company

Nov 1 Sandford Fleming given the honour of transmitting the first telegram on the completed All-Red-Route between Great Britain to Australia and New Zealand

1903 CPR creates Atlantic Telegraph Division from Megantic to Atlantic

* Donald Murray refines multiplex system

1905 Aug CPR enters long term agreement with the Temiskaming and Northern Ontario Railway for telegraphic rights and uses

1906 Lee DeForest invents first three-element vacuum tube

. Eastern Division divided forming Lake Superior Division from Chalk River to Port Arthur . Ontario Division extended to include Smiths Falls, Windsor and

Mactier to Hamilton . Western Division formed with part of Saskatchewan & Alberta

1907 CPTelegraphs attempts to quadruple prices charged to three Winnipeg newspapers. They object so CPR cuts them off plus one in Nelson B.C.

1909 Oct 28 CPR begins using telephone for train dispatching

1910 Central Division formed and divided into two parts (1) Saskatchewan District and (2) Alberta District . Board of Railway Commissioners of Canada established. It is designed to watch telegraph-telephone messages and fees charged for transmissions . CPR loses to Board over 1907 decision to raise prices . CPR drops news service

1911 CPR leases wire to The Canadian Press

1912 Aug CPR installs first Morkum Printer between Montreal-Toronto

1913 CPR installs other Morkum Printer's between Montreal-Winnipeg-Ottawa

1915 Jan 1 Canadian Northern Telegraph Company acquires the GNWTCo by agreement with Western Union

1917 CPR enters long term lease with The Canadian Daily Press reaching all across Canada / enters agreement with the SOO Line to install wires from Sault St. Marie Michigan to Emerson, Manitoba

1918 Dominion government takes over Canadian Northern Railway Company and GNWTCo

1919 All-Red Route established when British Government secures the Imperial Atlantic Cable Company then cuts connection with New York and Emden, Germany then diverting cable to Halifax connecting directly with London, England

Jun 6 Canadian National Railways incorporates

1920 CPR installs Morkum Printer Montreal-St. John

Nov 1 Grand Trunk Pacific Railway existence ends

1921 CPR installs Morkum Printer Montreal-Halifax

Jan 1 Canadian National Telegraph Company established

1922 Jan 1 Canadian National Telegraph Company now includes Grand Trunk Pacific, Grand Trunk, GNWTCo under corporate title "Canadian National Telegraph Company"

May CPR installs Western Electric four-channel printer

1923 Jan 30 Grand Trunk Railway becomes an official part of the CNR ceasing to exist on this date

1924 Sep 1 Canadian National obtains specific Western Union property in British Columbia

1925 CPR installs Morkum Printer Montreal-Vancouver

Jan 12 Dominion Telegraph Securities Limited establishes holding company

Jan 15 Agreement between Canadian National Telegraph, Dominion Telegraph, American Union Telegraph, Western Union, GNWTCo and Dominion Securities Limited reached

1926 CPR installs Morkum Printer Montreal-Calgary

Jun CPR installs Drop-Printer

Year	Date	Event
1927		CPR enters second long term telegraph agreement with the SOO Line from Sault St. Marie to Noyes Michigan via Minneapolis
	Apr 26	CNR inaugurates Carrier Current between Montreal-Toronto
	Nov	CNR installs Carrier Current Toronto-Winnipeg
	Dec	CNR installs Carrier Current Winnipeg-Saskatoon
1928	Jan	CNR installs Carrier Current Saskatoon-Edmonton
	Jan 1	CNR consolidates all telegraph accounts under Canadian National Railways
	Oct 2	CPR installs Carrier Current Montreal-Toronto
1929	Jan 4	CNR purchases Western Union and Anglo-American Telegraph Company rights in the Maritimes
	Mar 14	CNR installs Carrier Current Edmonton-Vancouver
	Jul 1	CNR obtains all Western Union and Anglo-American Telegraph Company property in the Maritimes
	Aug	CNR installs Carrier Current Toronto-Montreal making it 20 channels
	Aug 1	CPR installs Teletype printer from Montreal to Chateau Frontenac Hotel
	Mar 25	CPR installs Carrier Current Toronto-Winnipeg
	Sep	CNR installs Carrier Current Winnipeg-Saskatoon upgrading to 10 channels
	Sep 30	CPR installs Carrier Current Winnipeg-Vancouver
1930	Apr 15	CPR installs Carrier Current Montreal-Winnipeg
	Dec 1	CPR installs Carrier Current Montreal-Toronto-Windsor
	May	CPR installs Start-Stop Printer
1931		CPR installs Carrier Current Toronto-Windsor
1932		CPR adopts new corporate name for the telegraph system becoming "Canadian Pacific Communications" . CPR and CNR secure a joint contract pooling telegraph facilities so Canadian Radio Commission (later CBC) can operate
	Mar 1	CPR installs Carrier Current Montreal-St. John
	Jun 1	CPR installs Carrier Current St. John-Halifax
1937		CPR introduces stock ticker service
1948	Nov 1	Canadian National obtains all properties and titles of the Great North Western Telegraph Company
1949	Apr 1	Canadian National takes over telegraph service in Newfoundland
1950		Telegraph use diminishing due to Telex, telephone, deskfax
	Mar 1	CNR purchases all facilities of Anglo-American Telegraph held by Western Union in Newfoundland
	Apr 3	Dominion Securities and Dominion Telegraph completely absorbed into Canadian National
	Dec 13	Anglo-American and Western Union agree to let Canadian National Telegraphs operate Newfoundland telegraph
1954	Jun 1	Canadian National obtains all remaining properties of Montreal Telegraph and Montreal Telegraph ceases to exist on this date
	Jun 8	CNR obtains all remaining properties of the Grand Trunk Pacific Telegraph Company
1956	Jun 11	Canadian National Telegraphs is a wholly owned company and subsidiary of the CNR
1960		CN Telegraphs now titled CN Telecommunications
1970		Telesat Canada launches Canadian telecommunication satellites

1972	May 30	Last official railway telegraph message sent from Batiscan, Quebec
1980		CN and CP Telecommunications merge
1982		CPR has 50 per-cent joint partnership with CN Telecommunications

APPENDIX ONE

COMPANY FOUNDERS AND BOARD OF DIRECTORS OF CANADA'S EARLY TELEGRAPH COMPANIES 1846 to 1881

TORONTO, HAMILTON, NIAGARA AND ST. CATHARINES ELECTRO-MAGNETIC COMPANY

* Founded - October 1846 *
Thomas D. Harris and Clarke Gamble (Canadian)
Samuel Porter, Manager (USA)
Percy B. Marling, Superintendent (USA)

MONTREAL AND TORONTO MAGNETIC TELEGRAPH COMPANY - amended to - MONTREAL TELEGRAPH COMPANY

Organized December 26, 1846
Incorporated July 28, 1847

*** 1847 First Committee ***

Andrew Shaw, President
George Elder Jr., Director
John Glass, Director

L.H. Holton
Fred A. Wilson
Thomas Tait
Henry Joseph
William Murra

John Young, Chairman
Henry Chapman, Director

D.L. MacPherson
T.W. Middleton
Henry Starnes
I. Taylor

*** 1851 Election Results ***

James Parsons, Manager
James Dakers, Secretary
H.P. Dwight

O.S.Woods, Superintendent
Hugh Allan, President

CANADA GRAND TRUNK TELEGRAPH COMPANY

Organized March 16, 1852
Incorporated December 14, 1852
Provincial Statute 16th Victoria, Chapter 10

George H. Cheney, President (Toronto)
Philip Low, First Vice-President (Picton)
William Workman, Second Vice-President (Montreal)
F.A. Whitney, Secretary and Treasurer (Toronto)
Larratt W. Smith, Solicitor (Toronto)
James R. Boyd, Superintendent (Toronto)

Larratt W. Smith
John T. Arnold
W.P. Howlandv
J. Lukin Robinson
M.W. Brown
G.J. Grange
W. Fraser
W. Weller
A. MacFaul
Cecil Mortimer
Thos. Kirkpatrick
Alfred Hooker
Robert Anderson
John Anderson

F.A. Whitney
A.K. Boomer
J.G. Joseph
Wm. Bellhouse
H.B. Bowman
James Patton
W. Cluxton
H. Bull
J.P. Downes
J.R. Forsyth
Jno. Miller
Hon. R. Matthieson
John Leeming
J.F. Bradshaw

DOMINION TELEGRAPH COMPANY

Initial Organization 1868
*** Founder ***
Selah Reeve

* Officers and Directors *
January 11, 1871 Reorganization
Incorporated April 14, 1871
Reign 34 Chapter 52

Hon. John McMurrich, President/Director
J.I. McKenzie, Vice President/Director
James Michie, Treasurer/Director
Hon. J.H. Cameron, Consulting Counsel
Hon. M.C. Cameron, Solicitor
H.J. Colles, Secretary

A.R. McMaster, Director
Hon. W. Cayley, Director
S. Nelson, Director

L. Moffatt, Director
A. Copp, Director
T.N. Gibbs, Director

GREAT NORTH WESTERN TELEGRAPH COMPANY

*** Provisional Directors ***
May 7, 1880 - Charter
Reign 43 Chapter 666

Hon. John Norquay
Hon. Alex. Morris
Hon. C.P. Brown
E.P. Leacock
C. MacDonald
Hon. Dr. Schultz
A. Markham
J. Anderson
G.A. Kirkpatrick
Hon. A.G. Bannatyne
J.S. Aikins
H.S. Donaldson
Thomas Swinyard
F.W. Strange
Jas. J. Foy

*** Elected Directors / Officials - August 30, 1881 ***

Erastus Wiman, President
O.S.Wood, Vice-President
H.P. Dwight, General Manager
Frederic Roper, Secretary(ret)
Arthur Cox, Treasurer
James Daker, Secretary

Hon. Wm. McDougall
G.D. Perry
James Hudley
A.S. Irving
W. Gooderham
D.H. Bates
Richard Fuller
Adam Brown

*** Superintendents - August 30, 1881 ***

R.F. Easson
Superintendent of Press and Commercial News, Toronto

W.S. Battin
Assistant to the General Manager, Toronto

J. Poustie
Superintendent of Construction and Maintenance, Montreal

Angus Grant, Montreal
B.B. Toyes, Toronto
Warren Dow, Plattsbury, New York
N.W. Bethune, Ottawa
Edwin Pope, Quebec
H. McDougall, Winnipeg

CANADIAN PACIFIC RAILWAY

February 15, 1881 Charter
Dominion Statutes Chapter 1

George Stephen, President
Duncan McIntyre, Vice-President

John S. Kennedy
James J. Hill
Pasco du P. Grenfell
C.R. Hosmer,
General Manager, Telegraphs

Richard B. Angus
H. Stafford Northcote
Charles D. Rose
J. de Reinach

APPENDIX TWO

1852

AN ACT TO PROVIDE BY ONE GENERAL LAW FOR THE INCORPORATION OF ELECTRIC TELEGRAPH COMPANIES

WHEREAS it is expedient to provide by one General Law for the Incorporation and regulation of Companies formed for the purpose of constructing lines of Electric Telegraph in this Province: Be it therefore enacted by the Queen's Most Excellent Majesty, by and with the advice and consent of the Legislative Council and of the Legislative Assembly of the Province of Canada, constituted and assembled by virtue of and under the authority of an Act passed in the Parliament of the United Kingdom of Great Britain and Ireland, and intituled, An Act to Re-unite the Provinces of Upper and Lower Canada, and for the Government of Canada, and it is hereby enacted by the authority of the same, That

I. Any number of persons, not less than three, may associate for the purpose of constructing a line or lines of Electric Telegraph, with branches leading to and from the same, from and to any point in this Province, upon the terms and conditions, and subject to the liabilities prescribed in this Act.

II. Such persons, under their hands and seals, shall make a Certificate which shall specify: 1st. The name assumed to distinguish such Association, and to be used in its dealings, and by which it may sue and be sued, and a designation of the line or lines of Telegraph to be constructed by such association, and the route or routes by which such lines are to pass;
2dly. The capital stock of such Association, and the number of shares into which the stock shall be divided, and any provision which may be made for increasing, the same, the names of the Shareholders, and the amount of stock held by each;
3dly. The period at which the said Association shall commence and terminate; 4thly. A copy of their Articles of Association.

And such certificate shall be acknowledged before a Notary, and the original or a copy thereof, certified by such Notary, shall be filed in the office of the Secretary of the Province.

III. Upon complying with the provisions of the last preceding section, such Association shall be and is hereby declared to be a Body Corporate by the name so as aforesaid to be designated in the said Certificate, and a copy of such certificate, duly certified by the Secretary of the Province, may be used as evidence in all Courts and places for and against such Association.

IV. Such Association shall have power to purchase, receive and hold and convey, such real estate and such only, as may be necessary for the convenient transaction of the business and for the effectually carrying on the operations of such Association, and may appoint such Directors, Officers, and Agents, and make such prudential Rules, Regulations and By-Laws, as may be necessary in the transaction of its business, not inconsistent with the laws of the Province.

V. Such Association is authorized to construct the lines of Telegraph designated in its Certificate, along any and upon any of the public roads and highways, or across any of the waters within this Province, by the erection of the necessary fixtures, including posts, piers or abutments for sustaining the cords or wires of such lines, provided the same shall not be so constructed as to incommode the public use of such roads or high-ways, or injuriously to interrupt the navigation of such waters, also upon any lands purchased by the Association; or, the right to carry their line over which shall have been conceded to them by the parties having a right to make such concession, nor shall anything herein contained be construed to confer on any such Association the right of building a bridge over any navigable water.

VI. Any person who shall wilfully and maliciously injure, molest or destroy any of the said lines, posts, piers or abutments, or the material or property belonging thereto, or in any way disturb the working of the said Lines of Telegraph, shall, on conviction thereof, be deemed guilty of misdemeanour, and be punished by a fine not exceeding ten pounds, or imprisonment not exceeding one month, or both, at the discretion of the Court before which the conviction shall be had.

VII. It shall be lawful for any Association of persons, incorporated under this Act, by their Articles of Association, to provide for an increase of their capital and of the number of their associates.

VIII. It shall not be lawful for any Association under this Act to contract debts exceeding one half of the amount of the Capital Stock of such Association; and all evidences of debt issued by such Association shall be issued and signed by the President and Treasurer thereof.

IX. Any Telegraph Association or Company now organized may become incorporated under this Act, on filing in the office of the Secretary of the Province a certificate authorized by a resolution of its Board of Directors, signed and certified by the Secretary of the Company, containing the particulars herein before required in like cases, and signifying its acceptance of this Act.

X. It shall be the duty of every such owner, Company or Association to transmit all despatches in the order in which they are received, under a penalty of not less than five nor exceeding twenty-five pounds, to be recovered, with costs of suit, by the person or persons whose despatch is postponed out of its order; except that any Message in relation to the Administration of Justice, arrest of criminals, the discovery or prevention of crime, and Government Messages or Despatches, shall always be transmitted in preference to any other Message or despatch, if required by persons connected with the Administration of Justice, and any person thereinto authorized by the Provincial Secretary.

XI. Any Operator of any Telegraph Line, or person employed by any Telegraph Company, divulging the contents of a private despatch, shall be deemed guilty of a misdemeanour, and on conviction shall be liable to a fine not exceeding twenty-five pounds, or to imprisonment for a period not exceeding three months, or both, in the discretion of the Court before which the conviction shall be had.

XII. Her Majesty may, at any time, upon giving one month's previous notice to the Company, assume possession of any such Telegraph line and of all things necessary to the sufficient working thereof, for any time to be mentioned in such notice, and may for the same time require the exclusive service of the Operators and other persons employed in working such line, and the Company shall give up possession thereof, and the Operators and other persons so employed shall, during such time, diligently and faithfully obey such orders, and transmit and receive such despatches as they may be required to receive and transmit by any duly authorized Officer of the Provincial Government under a penalty not exceeding twenty-five pounds for any refusal or neglect to comply with the requirements of this action to be recovered by the Crown for the public uses of the Province, with costs, in any way in which debts of like amount are recoverable by the Crown.

XIII. Her Majesty may, at any time after the commencement of any Telegraph line under this Act, and after two months' notice to the Company, assume the possession and property thereof, and upon such assumption, such line and all the property, real or personal, essential to the working thereof, and all the rights and privileges of the Company, as regards such Line, shall be vested in the Crown.

XIV. If any difference shall arise between the Company and those who act for the Crown, as to the compensation which ought to be paid to the Company, for any Telegraph line and appurtenances taken under the thirteenth section of this Act, or for the temporary exclusive use thereof under the twelfth section, such difference shall be referred to three Arbitrators, one to be appointed on the part of the Crown, another by the Company, and the third by the two so appointed Arbitrators, and the award of any two of the said Arbitrators shall be final; and in case of refusal or neglect by the Company to appoint an Arbitrator on their behalf, or if the two Arbitrators cannot agree upon a third Arbitrator, then such Arbitrator shall be appointed by any two Judges of the Queen's Bench or Common Pleas, in Upper Canada, or of the Superior Court in Lower Canada, on application on the part of the Crown.

XV. It shall be also lawful for any Municipal Corporation in this Province, or for any Joint Stock Company incorporated under any Act of the Parliament of this Province, to subscribe for and hold Stock in any Company to be formed under this Act, and to pay the amount of such subscription out of any Municipal or other funds not specially appropriated to any other purpose, and to levy money by rates for paying any such subscription; and such Municipal Corporation shall have such rights as a Member of the Company, and shall vote upon the Stock held by it in such manner and by the intervention of such person or officer, as shall be determined by the articles of Association.

[Archives of Ontario, Toronto]

APPENDIX THREE

GRAND TRUNK RAILWAY RULES - 1864

INSTRUCTIONS
FOR
TRACK REPAIRERS
WITH REGARD TO
THE TELEGRAPH LINES

1. The foreman of each gang, will be responsible for the good and efficient state of the TELEGRAPH LINES along the length of road to which he is appointed.

2. The aforesaid foreman shall, every day or whenever inspecting his length of road, CAREFULLY observe the telegraphs; and whenever they are broken, in contact with each other, or in any way out of order, he shall IMMEDIATELY repair them.

3. The said foreman shall also, on being notified by the operator in charge of any telegraph station, of a break or interruption on the lines, immediately send some of his gang, or go himself, to repair it; and should the break or interruption not be found on his division of road either send out the men of the next division, or go on until it is found, or the repairers from the opposite direction are met with. For the neglect of this duty, or any part thereof, he will be fined or dismissed.

4. Each gang of men must be provided with the necessary tools and materials for repairing, consisting, of one pair of telegraph spurs, one pair of clamps, one pair of plyers, and one file; also a supply of telegraph wire and insulators or caps, to be had on application to the telegraph superintendent.

5. The duty of the repairers, in addition to the foregoing, will be as follows:

To straighten and brace up all poles that are leaning, or in danger of falling.

To replace broken or burnt poles with new ones, which must be cedar, and peeled.

To keep the lines from touching trees, branches of trees, or buildings.

To keep asunder the wire, wherever there may be more than one on the same poles, by keeping the top wire tight.

To take charge of all spare telegraph poles scattered along the road.

To report to the Operator at the nearest telegraph station, any repairs they may make, and the cause, nature, and place of the derangement; and

In the absence of the foreman, to take the necessary steps to repair the telegraph, when ordered to do so by an operator in charge of a station.

6. The BRIDGE-MEN are enjoined to be particularly watchful of the telegraph where it crosses streams, and to use every effort to keep it in proper order. They must also report to the operator at the nearest telegraph station, any derangement of the wires or masts of their bridges.

INSTRUCTIONS FOR REPAIRING

1. When the line is broken, the wire should be loosened from one or two of the poles on each side of the break, by taking off the tie-wire. The two ends of the broken wire should then be screwed in the clamps, drawn together, and joined by a tight, compact splice. If the wires are too short to come

together, a piece of the same size must be put in.

2. After the joint is made; fasten the line up with the tie-wire to the insulator on the pole as it was at first, care being taken to turn the loose ends or points of the tie-wire, where wooden insulators are used, downward; but where glass insulators are used upward, and these ends should be as short as possible. Very little sag should be allowed in the line between the poles.

3. To make a proper joint in the wire, hold the two ends lapping each other, in the plyers; and then take five or six short turns with each end around the opposite wire. Samples of such joints can be seen in many parts of the line.

4. When the lines are found to be broken in several different places, near together, so that it will require much time to repair them properly, the breaks should be hastily united, and the lines then hung along the fence or trees, but clear of each other and of the ground, snow or rails, so that they may be worked while proper repairs are being made, which, however, must be made as quickly as possible.

5. When the wire is off several poles together, but not broken, it should be lifted and kept clear of the ground, and the other wires where there are two or more wires on the poles while it is being fastened up; communications can then be sent over it while the repairs are going on.

6. When insulators or caps are off any of the poles, they must be immediately replaced, and the wire refastened as before.

7. When the line requires repairing at a curve, care must be taken to fasten the wire on the outside of the curve, so that the strain may be against the post and not from it.

USE OF THE TOOLS.

The spurs are used in climbing the poles. They should be strapped on the leg so that the prongs protrude from the hollow of the foot.

The clamps are for the purpose of drawing together the wire when broken, and are used as follows:

Affix one clamp to each of the broken wires by means of the screws, leaving enough of the ends of the wire clear of the clamps, to make a joint of. Then pass the cord, which is attached to one of the clamps, around the wheel of the other, and draw them together. A joint can then easily be made with the slack ends of the wire.

Every foreman is required to have a copy of these rules in his possession, under pain of the fine of $1.00, or dismissal.

(Courtesy of Art Clowes, Rail and Transit, Upper Canada Railway Society)

APPENDIX FOUR

DOMINION OF CANADA STATUTES for TELEGRAPH, TELEPHONE AND CABLE COMPANIES

Compiled Listing of Railway Legislation 1867 to 1905 (No Legislation in 1887, 1889 or 1891)

REIGN	CHAPTER	COMPANY	YEAR
62-63	50	ALGOMA CENTRAL RAILWAY	1899
46	78	AMERICAN, BRITISH AND CONTINENTAL CABLE COMPANY, LIMITED, (The). Changed name to EUROPEAN AMERICAN, CANADIAN AND ASIATIC CABLE CO. LIMITED	1883
42	5	ANTICOSTI AND MAGDALEN ISLANDS, subsidy for telegraphic communication	1879
56	73	AUTOMATIC TELEPHONE AND ELECTRIC COMPANY OF CANADA, incorporated [actual year is 1869]	1839 [sic]
44	33	ASIA, Marine Telegraph to, providing for incorporation of a Company	1881
45	54	. Authority to issue Letters Patent to a Company extended	1882
43	67	BELL TELEPHONE COMPANY OF CANADA, incorporated	1880
45	95	. power to manufacture instruments; to issue bonds and to extend lines	1882
47	88	. Authority to increase capital stock	1884
55-56	67	. As to increase of capital stock and and limit of bond issue	1892
57-58	108	. Limitation of bond issue	1894
2 Edw.VII	41	. Increase of capital, and regulation of rates	1902
44	33	CANADA AND ASIA, telegraphic communication	1881
36	96	CANADA ATLANTIC CABLE CO., incorporated	1873
40	69	. Corporate powers amended, and time for completion of works extended	1877
53	98	CANADA CABLE CO., incorporated	1890
45	94	CANADA MUMUAL [sic] TELEGRAPH CO., incor. [actually the Canada MUTUAL Telegraph]	1882
32-33	63	CANADIAN AND EUROPEAN TELEGRAPH CO., incorporated	1869
35	96	. Time for commencement and completion extended	1872

REIGN	CHAPTER	COMPANY	YEAR
37	83	CANADIAN AND GREAT NORTHERN TELEGRAPH CO., incorporated	1874
2 Edw.VII	51	CANADIAN NORTHERN TELEGRAPH COMPANY, incorporated	1902
46	79	CANADIAN RAPID TELEGRAPH CO., incorporated	1883
3 Edw.VII	100	CANADIAN TELEPHONE AND TELEGRAPH CO., incorporated	1903
37	84	CENTRAL CANADA TELEGRAPH CO., incorporated	1874
60	14	COLUMBIA TELEPHONE AND TELEGRAPH CO., incorporated	1896
47	87	COMMERCIAL CABLE CO., incorporated	1884
43	65	COMPAGNIE (LA) FRANCAISE DU TELEGRAPHE DE PARIS A NEW YORK, certain powers granted to	1880
61	100	DAWSON CITY AND VICTORIA TELEGRAPH CO., incorporated	1898
34	52	DOMINION TELEGRAPH CO., incorporated	1871
37	83	. Time for commencing and completing extended	1874
42	68	. Powers extended to all parts of the Dominion	1879
		. See GREAT NORTH-WESTERN TELEGRAPH CO.	
ELECTRIC TELEGRAPH COMPANIES ACT, (THE), R.S.C., C. 132			(1886)
32-33	14	ELECTRIC TELEGRAPHS, as to under what circumstances a company may diverge from its original line	1869
38	26	ELECTRIC TELEGRAPHS, Marine	1875
44	50	EUROPEAN, AMERICAN, CANADIAN AND ASIATIC CABLE CO. (Limited), incorporated	1881
46	78	. Rights and liabilities continued; section 3 amended, and name changed to "The American, British and Continental Cable Co. (Limited)	1883
4-5 Edw.VII	93	FESSENDEN WIRELESS TELEGRAPH COMPANY OF CANADA, incorporated	1905
48-49	92	FORT MACLEOD RANCH TELEGRAPH COMPANY, incorporated	1885

REIGN	CHAPTER	COMPANY	YEAR
43	666	GREAT NORTH-WESTERN TELEGRAPH CO. OF CANADA, incorporated	1880
3 Edw.VII	149	MARCONI WIRELESS TELEGRAPH CO., incorporated	1903
35	95	MONTREAL TELEGRAPH CO., corporate powers extended to whole Dominion; capital stock increased	1872
36	95	. Powers extended as in above Act	1873
43	64	. Power to enter into arrangements with other companies; and as to wires underground and telephone wires	1880
45	93	. Acts consolidated and powers enlarged	1882
		. See GREAT NORTH-WESTERN TELEGRAPH CO.	
49	94	NORTH AMERICAN TELEGRAPH CO. (THE) incorporated	1886
4 Edw.VII	104	. Increase of capital authorized; head office to be at Winnipeg	1904
61	111	NORTHERN COMMERCIAL TELEGRAPH CO., incorporated	1898
61-63	120	. Authority to construct branches and to increase capita	1899
63-64	110	. Authority to increase capital	1900
4-5 Ed.VII	136	NORTHWEST TELEPHONE & TELEGRAPH CO., incorporated	1905
51	100	NOVA SCOTIA TELEPHONE CO., corporate powers in Nova Scotia and New Brunswick confirmed	1888
62-63	3	PACIFIC (THE) CABLE ACT, 1899, To provide for the establishment of direct sub-Marine telegraphic communication between Canada and Australia	1899
1 Edw.VII	5	. Repealing section 3 of the above Act, and guaranteeing payment of Canada's share	1901
32-33	14	TELEGRAPH COMPANIES (ELECTRIC), as to under what circumstances company may diverge from its original line	1869
38	26	TELEGRAPHS, MARINE ELECTRIC	1875
44	33	TELEGRAPHS, Marine, between Canada and Asia	1882
46	45	TELEGRAPH, Marine, Pacific Coast to Asia	1883
44	46	TELEGRAPH OPERATORS ON GOVERNMENT LINES, declaration as to secrecy	1881

REIGN	CHAPTER	COMPANY	YEAR
3 Edw.VII	70	TELEGRAPH SERVICE, To assist in establishing and maintaining telegraphic service for Canadian newspapers	1903
35	97	THUNDER BAY SILVER MINES TELEGRAPH CO., incorporated	1872
42	16	UNION ATLANTIC CABLE CO., incorporated	1877
4-5 Edw.VII	49	WIRELESS TELEGRAPHY IN CANADA, Act to provide for the regulation of	1905
62-63	131	YALE-KOOTENAY TELEGRAPH COMPANY, (LIMITED), incorporated	1899

APPENDIX FIVE

TELEGRAPH POLE AND WIRE MILEAGES FOR MAJOR TELEGRAPH COMPANIES

TELEGRAPH COMPANY	YEAR	POLE MILEAGE	WIRE MILEAGE
Toronto, Hamilton, Niagara and St.Catharines Electro-Magnetic	1846	89	222*
Montreal Telegraph	1847	540	842
	1870	7800	12147
	1875	12820	20000
	1877	13078	20402
	1881	12694	21607
Dominion Telegraph	1877	4983	782
	1881/2	5000	12000
Great North Western	1881	17694	33607
	1884	22507	44058
	1902	17838	35721
Canadian National	1922	21500	50000
	1927	23000	58600*
	1943	n/a	78100
	1945	22800	73000**
	1949	24000	86700**
	1970	13884	n/a
Canadian Pacific	1882	181	(1 wire)
		714.5	(2 wire)
	1883	897	n/a
	1886/7	4525	14508
	1900	6309	37112
	1902	9736	41354
	1915	14633	105937**
	1928	16391	160287**
	1936/7	17596	173341**
Western Union	1877 (Nova Scotia/New Brunswick)	4452	6989
	1877 (Prince Edward Island)	50	248*

* - estimated mileage from vague or conflicting sources
** - combined telegraph, telephone and cable
n/a - not available or no record

APPENDIX SIX

LAST SIGNIFICANT STATIONS CONSTRUCTED BY MONTREAL TELEGRAPH COMPANY

Matane to Fox River, Quebec
July 25, 1878

Matane
St. Felicite
Cap Chatte
St. Anne des Monts
Martin River Lighthouse
Cap de la Madeleine
Grand Vallee
Chlorodrome
Grand Etang
Fox River

APPENDIX SEVEN
CANADIAN TELEGRAPHIC 'FIRSTS'

October 1846
Toronto, Hamilton, Niagara and St. Catharines
Electro-Magnetic Company
Canada's First telegraph company

July 28, 1847
British North American Telegraph Company
Commercial and News Despatch Service
under Robert F. Easson
"The Father of News Gathering in Canada"
by telegraph

1851
F.N. Gisbourne completes
first telegraphic connection between
St. John's and Conception Bay, Newfoundland

1852
Montreal Telegraph Company
First to use General Laws for Electric Telegraph Companies
to acquire the
Toronto, Hamilton, Niagara and St. Catharines
Electro-Magnetic Company

November 22, 1852
F.N. Gisbourne
completes North America's first submarine cable
between New Brunswick and Prince Edward Island

1853
Grand Trunk Railway of Canada
initiates Train Dispatching by Telegraph

1854
F.N. Gisbourne
develops and sells
Trans-Atlantic submarine cable plan

July 1, 1867
Hamilton Office of the Montreal Telegraph Company
was the first Canadian office to have
electricity for lighting.
George Blake, Manager of the office
also put up Christmas lighting displays
with carbon pencils and lights

1878
Sandford Fleming
develops Standard Time Zones to
assist train timetables and telegram Rates

1896
Canadian Pacific Telegraphs
the first telegraph company to use
electrical motor generators in Ottawa terminal,
replacing batteries as main source of power
on telegraph circuits

1898
Canadian Pacific Railway
develops strengthened and improved copper wire;
begins replacing iron galvanized wire with copper
on mainline circuits between Toronto and Vancouver

November 1, 1902
Canadian Pacific Railway
All-Red Route Telegraph Circuit
of the British Empire Completed
(First Suggested 1848)

1924
Ellwood Wilson
pioneered Canada's Fairchild Aerial Photographic Survey
technique in 1924 while surveying route for
the Abitibi Southern Railway and communication lines

April 26, 1927
Canadian National Railways
assists in development of Carrier Current
then installed across CNR system

1932
Joint pooling of telegraphic services
between the CNR and CPR

May 30, 1972
Canadian Pacific Railway
last to use Telegraph to send message
in Canada

APPENDIX EIGHT

VARIOUS STYLES OF TELEGRAPH SIGNS USED BY CANADIAN NATIONAL AND CANADIAN PACIFIC ON THEIR RAILWAY STATIONS

Weston Station, CPR, ca1900. This photograph details several telegraphic points. Of immediate interest are the style and prominence of the telegraph sign common about this time (Style Four D). The pole height over the tracks had to be eighteen feet plus. The station pole is located behind the station. Note also the arrangement of crossarms for the single circuit and the parallelling pole lengths on either side of the tracks.
(Courtesy Canadian Pacific Limited, Image No. 23484, Photographer J.W. Heckman)

Grand Trunk Station at St. Williams, Ontario. Of interest is the Great North Western telegraph sign (Style One) at the Waiting Room entrance.
(CNR Public Affairs #X44977)

CPR, Richmond, Quebec, ca1880. The photo collectively deals with the CPR signs found, typically, at many stations. Note the CPR Telegraphy sign style, Dominion Express, a Lunch Room and Restaurant, a Dining Room; the train order signal, locomotive features and distant van.
(National Archives of Canada, PA 187611, Edward Locke Collection)

TELEGRAPH SIGN NOTES

SIGN	TELEGRAPH COMPANY	DATES	NOTES
1	Great North Western	1881-1921	. may appear as: G.N.W. & Cable Office . white letters on black metal or wood background . use of & common for era
2a	Canadian National	1920-1930	. line common / may have spears on each end or other type of graphic . & not as common
2b	Canadian National	1921-1930	. straight line may appear under TELEGRAPH . no ampersand (&)
2c	Canadian National	1925-1950s	. more common in 1950s is tin sign with raised letters on a blue background
2d	Canadian National	rare	. wood background - likely used as a replacement only
3	Canadian Pacific	1883-1900	. dividing line may appear above TELEGRAPH
4a	Canadian Pacific	1900-30s	. may also see RLY. . most frequent style
4b	Canadian Pacific	1900-30s	. frequent style
4c	Canadian Pacific	1900-30s	. frequent style
4d	Canadian Pacific	1900-30s	. less frequent style / may appear with or without 'OFFICE' on bottom
4e	Canadian Pacific	1900-30s	. less frequent / may appear with or without 'COMPANY'
4f	Canadian Pacific	1900-30s	. rare style
5a	Canadian Pacific	1929-60s	. various shades of blue, maroon and black/grey backgrounds /letters tend to gold, yellow and white . frequent style
5b	Canadian Pacific	1929-60s	. most frequent style
5c	Canadian Pacific	1929-60s	. frequent
5d	Canadian Pacific	1929-60s	. frequent

NOTE: Telegraph station signs were not exactly a priority for B & B crews or for that matter, the railway or the telegraph company unless it was in a highly populated area. Sign repair was rare unless a visiting dignitary or a superintendent complained of the appearance.

The following designs are representative of most styles with many variations on a theme. Signs were consistently in block capital letters and properly spaced on the board, most often wood, tin or other metallic background (less common). Dates are categorized on a general basis and do not represent specific style changes; the CPR Style (3) was seen along the CPR's Ontario and Quebec Railway stations as late as 1960.

SIGN

1

GREAT NORTH WESTERN
TELEGRAPH
& CABLE OFFICE

2(a)

CANADIAN NATIONAL
TELEGRAPH
& CABLE OFFICE

2(b)

CANADIAN NATIONAL
TELEGRAPH
AND
CABLE OFFICE

2(c)

CANADIAN NATIONAL
TELEGRAPHS

2(d)

CANADIAN
NATIONAL
TELEGRAPH
& CABLE OFFICE

SIGN

3

CANADIAN PACIFIC RY.
TELEGRAPH

4(a)

CANADIAN PACIFIC RY.
TELEGRAPH

4(b)

CANADIAN PACIFIC RY.
TELEGRAPH
COMPANY

4(c)

C.P.R.
TELEGRAPH
COMPANY

4(d)

C.P.R.
TELEGRAPH

4(e)

C.P.R.
TELEGRAPH
COMPANY OFFICE

SIGN

4(f)	C.P.R. TELEGRAPH & EXPRESS
5(a)	CANADIAN PACIFIC TELEGRAPHS
5(b)	CANADIAN PACIFIC TELEGRAPH
5(c)	C.P.R. TELEGRAPH CO.
5(d)	C.P.R. TELEGRAPH

GLOSSARY

AGENT - OPERATOR'S TELEGRAPHIG GLOSSARY

TERMS, CODES, SYMBOLS and COLLOQUIAL EXPRESSIONS

During the 1880s railways across North America began exchanging privileges and services. Services included a no charge user's fee for the railway poles, wires, dispatching of trains, and handling railway business and correspondence. Other privileges included transporting company employees at reduced rates or free on passenger trains. A special privilege was the free handling of telegraphic messages.

Colloquial language developed as interchange traffic expanded between railway companies throughout North America. Boomers moving from one railway to the next often carried codes and phrases not used at their new location. In time, others picked up and spread these new phrases. Colloquial terms, often descriptive and spirited, helped to spread understanding and lessen crossborder traffic confusion. A common working language meant greater acceptance of information and increased speed.

An example of a colloquial message keyed in Morse between operators might be: "Bn ot a nite smtg mst b dn wi th biz rt awa bt dt a aytg abt it pls. U kn hw tis urslf. Ha Ha"

Which decodes as: "Been out all night something must be done with this business right away but don't say anything about it please. You know how it is yourself (laughter)."

Note the lack of vowels where possible and must not be confused with Phillips Code. Phonetics played a large role in the following abbreviations. Many are extremely unique and interesting.

FOOTNOTE EXPLANATIONS and EXAMPLES

(A) 'sending in the air' - lower case between apostrophe's a colloquial expression;

(B) 'shipper' - lower case between apostrophe's a colloquial
'breezer' expression - top term most commonly used;
second line less common in use;

(C) BX - capital letters - official designated code/symbol for message use; capital letters used in the explanation indicate the code abbreviation origin
(eg) DS DiSpatcher;

(D) 92 - number only - official designated number for message use;

(D1) 77 (1) - number with footnote in brackets for known source
- if no footnote given, many sources with unclear origin(s) but likely social;

FOOTNOTE SOURCES:

(1) Grand Trunk Railway Rule Book 1911
(2) Canadian National and Canadian Pacific Telegraph Companies
(3) Unique to Canadian National Telegraph Company
(4) Unique to Canadian Pacific Telegraph Company
(5) Great North Western Telegraph Company
(6) Uniform Code of Operating Rules (various)
(7) various American sources

- compiled by the author

CATEGORY ONE: AGENTS, DISPATCHERS and OPERATORS

Most sent numeric codes and block letter symbols were for official transmission of telegraphic messages. They would be used in railway service, commercial messages, between operators late at night, or press wires.

1	Wait a minute. (1)
10	Keep your key closed. (Used by Chief Dispatcher and for testing operators (1))
12	How do you understand? (1)
13	I/we understand.(1) (Some sources: Understand? (7))
134	Who is it? Who is the sender? Who is at the key?
15	Weather report. (7)
17	Danger signal, to prevent an accident. Attend at once to danger. (7)
18	What is the trouble? (1)
19	Train Order (6)
2	Very important (1) -or- What's the time? (7)
21	Deliver immediately and get an answer. (1)
22	Busy on another wire. (1) You are breaking up, adjust your equipment.
23	Telegram for all offices. -or- Attention all operators. -and- All attend and take this telegram. (1)
25	I am busy. -or- Busy on all wires. (1)
27	Very important message. Priority message coming through. -or- Adjust your instrument. (7)
3	What time is it?
30	The end. No more. ((7) '92 Code'-1859)
31	Train Order
4	Where shall I go ahead (begin)? (1)
44	Answer quickly. (1)
5	Have you any thing (business) for me? (1) -or- Message for all. (7) -or- Close your key. (7)
55	Important (1)
6	I am ready. (2)

7	Are you ready? (2) Important. I have business for you. (7)
73*	My love to you. (Telegraphic Review-1857) Accept my compliments. ((7) '92 Code'-1859) Best regards. Best wishes. -or- My compliments. (2) * James Douglas Reid, born in Edinburgh, Scotland, March 22, 1819, emigrated to Toronto, then New York; was a close friend of Morse. It is believed Reid developed this numeric code, originally used as a symbol of fraternity among railway telegraphers.
77	Telegram for you. (1) Are you ready to copy/start?
8	Close your key and stop breaking (and adjust). (1) Busy on another circuit. (7)
82	Morning signal for yard reports in terminal areas.
88	Love and kisses. (Signed at the end of telegrams generally received/sent during World War II.)
9	Close your key - priority business. Clear the line. Signal for Train Orders. (1) Wire Test. (7) / Wire Chief Calling (7)
92	Deliver immediately. (1)
93	President's, Vice-President's, General Manager's, General Transportation Manager's business signal for their use only. (1)
94	Superintendent's business only signal. (1)
95	President's signal. (2) / Following very urgent (7)
A	Arrived (1)
ADR	ADdRess
AGN	AGaiN
AGT	AGenT (1)
AHR	AnotHeR (7) (-or- HR (7))
AM	morning
ANS	ANSwer (7)
B	Bulletins and Train Register Station (6)
BAT	BATtery (7)
BC	Block Clear (6)

BF	BeFore (7)
BK	equal sign -or- double dash -or- Stop sending -or- break (2) -or- back (7)
BKR	BreaKeR
BN	BeeN (7)
BTN	BeTweeN (7)
BTR	BeTteR
BX	squared brackets
C	Cents sign (2) -or- Coaling station (1) -or- yes
CFO	Cancel Former Order (1)
CK	ChecK
CCT CKT	CirCuiT (7)
CL	CalL -or- closing station
'closing out' 'signing off'	discontinue operation on a channel of communication (2)
'clear board'	no orders for approaching train / proceed
CO	COmpany
COL	COLlect (7)
COM	COMplete message (6)
CT	CannoT (7)
CWT	hundred WeighT (C = Roman numeral for 100)
CX	capital letter -or- lower case
CY	CopY message (6)
D	Day telegraph station (1) -or- English pence (4)
DH	DeadHead -or- free message (using a frank) (1)
DN	Repeat when seNding -or- DowN
DOT	Decimal pOinT
'double hoop the train'	agent must hoop train orders on two separate hoops to a double header steam train / each crew must receive the same orders / also for triple headers 'triple hoop...'

DS	DiSpatcher (1)
DT	Do noT (7)
DX	dash (2) -or- distance (7)
E	fraction line (2)
E5	note code (2)
EM	EMpty (2)
ENG	ENGine (6)
EX	EXtra train without a schedule (1) -or- EXpress (7)
EXCN	EXCursioN train
f	Flag station stop (6)
'falls down'	makes or making an error in transmission
'fill in'	message must be retransmitted
FRT	FReighT train (6)
FM	FroM (-or- sometimes FRom FR)
GA	Go Ahead and send -or- Good Afternoon
GBA	Give Better Address (7)
GB	Good Bye (also Phillips Code)
GD	GooD (also Phillips Code)
GE	Good Evening (also Phillips Code)
GG	GoinG (7)
GM	Good Morning (also Phillips Code)
GN	Good Night (also Phillips Code)
GPM	Good afternoon
HA	signifies a laugh
HHD	HogsHeaD (or locomotive engineer)
HI	rare - used as 'telegraphic laughter' (found only one occurrence)
HM	HoMe
HX	hyphen (2)

I5	small letter -or- lower case (2)
IMMY	IMMediatelY (7)
IMPT	IMPortanT
JCT	JunCTion (6)
K	standard clock, bulletins and train order station (6)
KO	colon (2)
KQ	colon quotation (2)
KW	KnoW
KX	colon dash (2)
L/LV/LVG	Leave station (1) Leave or LeaVinG (1) and (7)
LX	English pound sterling sign (4)
MIN	MINutes (6)
MM	paragraph (3)
MN	exclamation point (3)
MSG	MeSsaGe (6)
MSGR	MeSsenGeR (7)
MST	MiSTake (1) -or- MSK (7)
N	Night telegraph station (1) -or- No (7)
NM	No More
NO	Number (6)
NS	No display of train order Signal (6)
NTG	NoThinG
NW	NoW
OB	Old Boy
OK	correct (2)
OPR	OPeRator (6) -or- OP
OM	Old Man
OS	train Order Sheet-a position report (6) or On Schedule Out of Station.

OSO	per cent sign (4)
OT	On Time (1) -or- Old Timer
P	telePhone -or- T (6)
PC	per cent sign (3)
PD	PaiD (7)
PKG	PacKaGe
'pick himself up'	to correct an error and get on with the work
PN	start PareNthesis (2)
PO	Post Office (7)
PSGR	PaSsenGeR (6)
'put a big S up' 'show a big S'	dispatcher orders agent to put a lower priority train into a siding so a first class train can pass (3)
'put the red board up'	operator displays a stop signal under the direction of the dispatcher so a train will receive its orders from that station
PX	boomer operator asking for the price of a meal. -or- sometimes refers to the Press
PY	end parenthesis (2)
QJ	end quotation (2)
QN	start quotation (2)
QK	QuicK (7)
QX	quotations within quotes (3) or apostrophe (4)
'rap' 'Rolling A Pill'	signal used by operator indicating he/she is rolling a cigarette-sender needs to hold the circuit open a few minutes
RCD	ReCeiveD
RIG	station equipment
RJ	relief Operator called so regular OP has a break
RPT	RePeaT your transmission (5)
RR	Repeat (1)
S	regular station Stop (6)
'say when'	report when the order has been carried out or delivered
SD	Signal Displayed adding an R or Y as required (6) -or-

	Stop Displayed (1)
SECN	SECtioN train following
'sending in the air (wind)'	sending a message but no operator is present to receive and copy it down
SI	semicolon (2)
SIG	SIGnature
SINE	give private or office signal. (7)
'slow board'	agent to indicate slow speed to approaching train who must pick up orders but not to stop
SS	SteamShip
STN	STatioN (6)
SUB	SUBdivision (6)
SW	SWitch (1)
SX	dollar(s) (2)
T/TEL/P	Telephone station (1) - also P (6)
'take a short'	obtain an relief operator (RJ) for a short rest
TFC	TraFfiC
TR	TRansmit
TT	ThaT
TU	Thank yoU
UJ	underline (end) (2)
UM	signifies a grunt (7)
UT	English shillings (4) -or- fraction line -or- diagonal (2)
UX	underline (2)
W	Water station (1)
WB	WayBill (2)
WT	WhaT
'wire'	circuit - also this signal pre-exempts a circuit and takes preference over everything else except a 95
'without turning a hair'	able to copy 30 messages in one hour without stopping (CNT)

X	train will be held until order is complete (1) -or- crossover between main tracks (6)
Y	wYe at this station (6)

CATEGORY TWO:
AROUND THE OFFICE

As with most offices workers, many phrases are shortened forms of words regularly repeated.

'blue slip'	incorrect message sent by an operator is returned to him for correction
'butterfly' 'hog trough'	resonator hood for sounder
'check'	performing a word count in a message for toll calculation
'clocked'	an error written on the message with the wrong time recorded
'copied'	a finished message that has been received or taken down
'copy'	copying down a message as it is being sent -or-an operator's copy of a sent telegram for record keeping purposes
'drag'	the normal time it takes to move the messages over the wire in an orderly fashion according to the Rules
'file'	messages entrusted to the telegraph company for sending
'filing'	presenting messages to the company for transmission
'follows copy'	sending exactly what is on a message form
'whiskers on it'	message that has been lying around longer than necessary -or- due to circuit failure, messages can not be transmitted
'an iron horse'	a transmitter enabling an operator to work at the same speed as the MUX teleprinter
'laid out'	delayed messages finally getting sent out on the circuit
'messages' 'traffic'	telegrams for transmitting
'mill'	typewriter (Henry Mill invents first one 1714) 'threshing machine'
'moved'	messages must be transmitted quickly
'on closed key' 'on the take'	receive messages without interruption
'on the hook'	an operator who likes a few extra messages unsent before him to always appear to be busy
'pile up'	an error where over-typing occurs at the receiving end because of a slow operator on a typewriter

‘piled up’ — overloaded with work -or- unable to keep up with a sender

‘pr’ — Personal Record Envelope where all the operator’s errors are recorded for Educational demerit points

‘prize package’ ‘stiff/torpedo/’ ‘valentine’ — a customer’s complaint and all correspondence which relates to that customer - generally a known troublesome customer

‘put on the wire’ ‘shoot’ — transmit messages

‘putting ten on a line’ — operating a typewriter to have ten words on each typed line to assist the operator in totalling the number words for toll charges or credit

‘run-over’ — overwhelmed with work -or- can not keep up with demands

‘running down’ — tracking down an error -or- locating an error

‘sheet’ — business orders

‘threshing machine’ — typewriter

CATEGORY THREE: CIRCUITS

Electrical wire circuits had to be in prime operating condition at all times. But failures, for whatever reason, did develop. Agents were responsible for batteries and regular maintenance to their key sets, with the more technical work left to the Engineering Department after 1900 or 1910.

‘average’ — the number of messages which are handled per average hour of business

‘city’ ‘city wire’ — local telegraph line generally to a nearby branch office

‘drops’ — intermediate connections

‘heavy’ — see ‘trunk line’

‘round robin’ — telegraph circuit starting and ends in the same city

‘tie line’ — local telegraph line to a customer in the immediate vicinity

‘trunk line’ — heavily loaded telegraph circuit generally between two (or more) large offices

‘way wire’ — telegraph line connected to several connections along a subdivision

CATEGORY FOUR: MALFUNCTIONS

These terms were used when complete electrical failures or interruptions caused a circuit to fail in whole or part.

'au'	printer tape inserted upside down and can not be read
'bug'	'something' wrong with the electrical connections (see also Operators)
'chatter'	a failed circuit
'copperplate'	manifold tissue paper interleaved with carbon paper
'flimsy'	tissue paper for Train Orders
'flop'	a failed circuit
'frozen'	electrical contacts are stuck, frozen or burnt together
'glass arm'	paralytic condition of the arm which many Bunnell operators experienced - less frequent with BUG or vibroflex operators
'hit'	a failed circuit meant time off until fixed
'hooked up all fooey'	wires crossed or not connected properly by the Engineering Department causing problems
'jar'	a failed circuit
'juice'	electrical current cutoff at a repeater station
'junk'	unintelligible letters/code sent by a teleprinter
'kick'	a failed circuit after experiencing an electrical surge
'like a bell' 'like a charm'	a perfect operating circuit
'line-up'	series of steps which must be taken in order to fix a failed circuit
'mux' 'mux operator'	a multiplex printer and the man who ran it
'run a-mux'	a teleprinter out of control -or- MUX operator
'out of sync' 'out of wack'	a circuit failure (SYNChronism)
'patched in'	splicing in a new section of wire and replacing a failed section
'scrape'	a failed circuit
'stick'	also pen or stylus - a writing utensil to write a flimsy or a stylus for several pages with carbon paper

‘swing’	a failed circuit often at night on the ‘swing’ shift
‘wobble’	a failed circuit where the code alters between fine and garbled

CATEGORY FIVE:
WIRE CONDITIONS

These are more specialized terms or phrases for the wire itself.

‘all failed’ ‘busted up’	wire or circuit is inoperative
‘blew up’	line is down or inoperative
‘dropped out’	dots and/or dashes are being left out from a light handed sender resulting in failures or errors in the message code - operator needs to adjust set
‘failures’	difficulty understanding the code of a light operator
‘getting nix’	no messages being sent or received as the line is down
‘heavy’	a heavy handed operator who has sluggish code or could also be the actual circuit
‘in the mud’	sluggish code from a ‘heavy’ operator
‘light’	when operators do not depress the key firmly enough creating improper code
‘pull it’	to take a circuit out of service for repairs or problems needing repair
‘rotten string’	bad circuit or wire because of various problems

CATEGORY SIX:
MESSAGE CODES, SERVICES, MESSAGE TYPES

Railway agents handled two types of telegraphic messages. The first was direct railway communication with the dispatcher, and other operators. Messages were relayed to areas outside a specific divisional point as required. The second type was commercial. Commercial messages were sent from small rural towns or villages and the agent was required to handle both message forms. The railway companies demanded that their operators co-operate in anyway possible in accepting telegrams. The administration believed that it fostered goodwill and encouraged patrons to use CN and CP’s telegraphic services.

‘blacks’	ordinary messages before Western Union made the telegram form yellow (7)
‘blue’ ‘blues’	Day letters (5)
‘book’	message intended for duplication and sent to an identical delivery but to a number of different addresses (4)
‘breezer’ ‘shipper’	name given to a messenger delivering a message a long distance from the telegram office

'bromidics' 'canned'	ready-made greetings for a customer (7)
'busts' 'busted'	spoiled copy in transmission and put into a dead file
Cablegram	submarine cable message
CDE	CoDE for cablegram (4)
CLT	Cable LeTter (7)
Code Rate	(cable)
Day Lettergram	full charge telegram during business hours
Deferred Rate	(cable)
DH	Dead Head - a free telegram or non-revenue message
DL	Day Letter (2)
'duped'	a message duplicated or re-sent
Fixed Text Message	sent during Yuletide season or other Holiday times - usually special form like Santa Claus - had a fixed tariff (2)
Full Rate	full charge of tariffs to a day telegram/cablegram
Full Rate Telegram	sent during day - full tariff
'gossip'	general financial news or other types of business orders, quotations
HTTC	'Hail To The Chief'
ICO	deferred telegram/cablegram (7)
'killed'	a filed message then withdrawn by sending customer
LC	deferred cablegram (4)
'mounties'	boys on bicycles who deliver telegrams or call boys
NIGHT	NIGHT message (5) or NITE = NIghT messagE (5)
Night Letter	sent during night - reduced tariff
Night Lettergram	(cable)
Night Message	accepted up to 2 A.M. / sent at night - reduced tariff (3)
NL	Night Letter-accepted to 2 A.M. / sent at night(5)
NLT	Night cable LeTter (4)

NM 'pink'	Night telgraM (2) rush telegram or caution card (7)
Post Lettergram	(cable)
Press Message	Scoopman -or- scoop wire
'query'	used by newspaper reporters to send in a story
'reds'	Night message (7)
Urgent Rate	mostly used by Brokers and financial houses (cable)
Wiregram	(cable - Wiregram)
WLT	Weekend LeTter

CATEGORY SEVEN: OPERATING DEPARTMENT

The Operating Department -later Communication Department- was responsible for the operation and maintenance of all railway communication lines and instruments. These responsibilities included the telegraph instruments, telephone dispatching circuits, teleprinter circuits, long distance telephone lines and even the public address systems in the large stations.

'blind'	one way communication and unable to acknowledge message
'blockite'	a boomer out of work and looking for handouts
'blow up'	quit a job suddenly especially boomers
'boomer'	itinerant Morse operator who moves from city to city for any available Morse work then moves on
'brass pounder' 'pounding brass'	original telegraph operators, operating a brass key set
'bug line'	telephone connection between the engine house and the yard office or telegraph office/station
'carrying a card'	operators who belong to the "Order of Railroad Telegraphers" union
'come along'	tool used to pull wire taut
'extra'	can not get work so put on an extra list and having to wait for work
'flag' 'flagged'	an alias or assumed name used by boomers who do not 'carry a card' or working illegally
'getting a sign'	an out of work boomer trying to get a job 'signing on'
'the graveyard'	night shift
'grunt'	a lineman's assistant
'home guard'	operator who stays at the same location or with the same railway and does not move on (opposite to boomer)

'hot rail'	linemen shout this when a train is approaching along the right-of-way while they work on the wire(s)
'old man'	term of affection applied to a head of the office such as the Wire Chief, Chief Operator, Traffic Manager
'on the block'	out of work, especially applied to a boomer
'on the wagon'	asking for a cup of coffee by an out of work boomer
'ops'	operator(s) (6)
'rj'	relief operator
'scoop'	working overtime at a press key telegraph office
'sign' 'sine'	letters assigned to identify an operator by 'old man' or one could choose the first letter as long as it was different from others in the area (eg) SM
'soak'	a boomer who pawns something to pay for a meal
'squared'	boomer who was provided with a free ride, meal or a place to sleep for the night
'trick'	shift time or working hours
'wolf'	someone who takes/receives more overtime work than most and appears to be 'hogging it'

CATEGORY EIGHT:
OPERATORS SKILL LEVEL and MISCELLANEOUS TERMS

Speed and accuracy were paramount to the successful operation of railway service correspondence and commercial messages. Many of the phrases below reflect a form of endearment or affection.

'bug'	Vibroflex telegraph key which operated for side-to-side. Also known as a 'sideswiper'. Generally 25 to 30 words per minute faster than a Bunnell key. Preferred by 'lightning bug' operators.
'burning up the wire'	hogging the circuits / a fast sender with a poor receiver
'electric owl'	night operator
'fast wire'	a high speed working circuit
'fist' 'mitt'	sending operator whose Morse was excellent and had a good constant touch and reliable code / also applied to their handwriting as many were quite beautiful
'get up'	an operator who needs a break and wants an RJ
'going over' 'taking the other side'	not capable of receiving code well
'ham'	poor telegrapher who is usually a student or beginner

'hanging up the clock'	selling or hocking your railway watch when leaving the company
'heavy'	tough, impatient operator
'hold down'	maintain a good sending speed
'lid'	see 'ham'
'lid operator'	working in the relay office covering local wires that go to all small line stations
'lifted'	an operator being relieved from long duty for a rest or an operator who is unable to maintain the pace and is taken off the circuit
'lightning bug' 'lightning slinger'	fast sending operator - prefer the Vibroflex key
'outbound'	a fast sending operator with a poor receiving operator at the opposite end
'pasted'	receiving operator who can not keep pace with a sender
'plug'	landline telegrapher
'roasting'	a receiving operator getting ridiculed from a fast sender
'seat hog'	what agents call a passenger who takes up more than one seat in a waiting room or coach while others are standing
'stuff'	messages easily read without guessing any part of a message
'under the flag'	a boomer who is working under his alias while being blacklisted as an operator so as not to be caught
'under the table'	a receiving operator who is swamped with messages and can not get it all written down
'working bonus'	an operator who gets premium wages depending on his/her output
'write'	operator to make some noise on the circuit, send figures or anything so that the wire can be adjusted while another operator is sending

FOOTNOTES:
CHAPTER ONE - CANADIAN NATIONAL RAILWAYS

1 Glazebrook, G.P. de T., A History of Transportation in Canada,
Vol.1, P.136
2 IBID., P.137

... refers to chapter title ...
The Toronto, Hamilton, Niagara and St. Catharines Electro-Magnetic Company was also known as: The Toronto, Hamilton, Buffalo Telegraph Company; the Toronto, Buffalo Electro-Magnetic Telegraph Company; the Toronto, Hamilton, and Niagara (Electro-) Magnetic Telegraph Company.

3 Encyclopedia of Canada, Volume VI, P.118
4 Easson, R.F., The Telegraph in Canada, P.3

... refers to page 3 chapter one ...
(Ellwood Wilson was the first to include telegraph, telephone, electrical power lines, forestry and mine sites on a map, pioneering "Canada's Fairchild Aerial Photographic Survey" technique in the spring of 1924; he surveyed a railway route for the Abitibi Southern Railway through the Laurentians, and the QNS & L later did the same).

5 Murray, J., A Story of the Telegraph, P.108
6 Easson, op. cit., P.4
7 Canada and Its provinces, Vol. 15, P.313
8 Warner, D.G., The First Fifty Years of the Canadian Telegraph, P.4
9 Easson, R.F., The Telegraph in Canada, P.4
10 Murray, J., A Story of the Telegraph, P.110
11 Canada Grand Trunk Telegraph Company, Minutes 1854, P.5
12 IBID., P.6
13 IBID., P.6
14 IBID., P.6
15 IBID., P.7
16 IBID., P.7
17 Easson, R.F., The Telegraph in Canada, P.5
18 IBID., P.5
19 Murray, J., A Story of the Telegraph, P.110
20 Warner, D.G., The First Fifty Years of the Canadian Telegraph, P.75
21 Coles Canadiana Collection, The Railways of Canada, P.61
22 IBID., P.61
23 Green, E., Canada's First Electric Telegraph, P.368
24 Cote, J.G., CNR Card Files, Card #P1048
25 Warner, D.G., The First Fifty Years of the Canadian Telegraph, P.6-7
26 IBID., P.5
27 CNR Corporate History, CNR DeChief Library, P.160
28 Canada Grand Trunk Telegraph Company, Minutes 1854, cover
29 Clowes, Art, Courtesy of, Upper Canada Railway Society
30 Murray, J., A Story of the Telegraph, P.110-111
31 Warner, D.G., The First Fifty Years of the Canadian Telegraph, P.8
32 CNR Corporate History, CNR DeChief Library, P.161
33 Cote, J.G., CNR Card Files, Card #S1511-12
34 CNR Corporate History, CNR DeChief Library, P.161
35 IBID., P.161
36 Cote, J.G., CNR Card Files, Card #1393
37 IBID., Card #P1102
38 Easson, R.F., The Telegraph in Canada, P.6

39 IBID., P.6
40 CNR Corporate History, CNR DeChief Library, P.158
41 Easson, R.F., op.cit., P.9
42 Murray, J., A Story of the Telegraph, P.118
43 IBID., P.118
44 Cote, J.G., CNR Card Files, Card #K1316
45 Easson, R.F., The Telegraph in Canada, P.7
46 CNR Corporate History, CNR DeChief Library, P.161
47 Hathaway, R.H., The Telegraph in Canada, P.2
48 CNR Corporate History, op.cit., P.165
49 Statutes for the Dominion of Canada, Chapter 51, 1902, no P.
50 CNR Corporate History, op.cit., P.166
51 IBID., P.166
52 IBID., P.159
53 IBID., P.166
54 IBID., P.166
55 Thornton, J.J., Keeping Pace with Prosperity, P.15
56 CNR Corporate History, op.cit., P.160
57 IBID., P.160
58 IBID., P.162
59 IBID., P.158
60 Glazebrook, G.P. de T., A History of Transportation in Canada, Vol.1, P.96
61 Warner, D.G., The First Fifty Years of the Canadian Telegraph, P.10
62 Lacombe, D.V., Celebrates Telegraph Centenary, P.14
63 IBID., P.14
64 Murray, J., A Story of the Telegraph, P.113
65 Warner, D.G., The First Fifty Years of the Canadian Telegraph, P.13
66 IBID., P.12
67 IBID., P.12
68 Woods, S.E., Cinders & Saltwater, P.27
69 Herstein, H.H., et al, Challenge and Survival:The History of Canada, P.221
70 MacKirdy, K.A., et al, Changing Perspectives in Canadian History, P.212
71 British North American Act 1867, Section 145
72 IBID., Section 92
73 Woods, S.E., Cinders & Saltwater, P.125
74 IBID., P.130
75 CNR Corporate History, CNR DeChief Library, P.164
76 IBID., P.164

CHAPTER TWO - CANADIAN PACIFIC RAILWAY

1 Innis, H.A., History of the Canadian Pacific Railway, P.38
2 IBID., P.39
3 IBID., P.40
4 Mackay, D., The Asian Dream, P.29
5 Stevens, G.R., A History of the Canadian National Railways, P.60
6 Mackay, D., op. cit., P.34
7 Innis, H.A., History of the Canadian Pacific Railway, P.41
8 Mackay, D., op. cit., P.34
9 IBID., P.34
10 Glazebrook, G.P. de T., A History of Transportation in Canada, Vol.2, P.41
11 Innis, H.A., History of the Canadian Pacific Railway, P.40N
12 Glazebrook, G.P. de T., op. cit., P.39
13 IBID., P.40
14 Innis, H.A., History of the Canadian Pacific Railway, P.41
15 IBID., P.42

16 IBID., P.41
17 Mackay, D., The Asian Dream, P.36-7
18 Glazebrook, G.P. de T., A History of Transportation in Canada,Vol.2, P.42
19 IBID., P.44
20 Mackay, D., The Asian Dream, P.37
21 Glazebrook, G.P. de T., op. cit., P.42
22 Mackay, D., op. cit., P.36
23 Mackay, D., op. cit., P.42
24 IBID., P.42
25 Innis, H.A., History of the Canadian Pacific Railway, P.83
26 IBID., P.85
27 IBID., P.85
28 IBID., P.85
29 IBID., P.85
30 IBID., P.86
31 Canadian Pacific Communication Department, Chronology, Pt.1, P.1
32 Canadian Pacific Communication Department, Doc.#277, P.1
33 Innis, H.A., History of the Canadian Pacific Railway, P.88
34 Canadian Pacific Communication Department, Chronology, Pt.1, P.2
35 Batten, J., Canada Moves Westward 1880-1890, P.38
36 Canadian Pacific Charter, Section 16, Feb. 15, 1881
37 Innis, H.A., History of the Canadian Pacific Railway, P.41
38 Canadian Pacific Facts and Figures (1937), P.41
39 Warner, D.G., The First Fifty Years of the Canadian Telegraph, P.20
40 Dots and Dashes, April-May-June 1977, P.9
41 Innis, H.A., History of the Canadian Pacific Railway, P.133
42 Canadian Pacific Communication Department, Chronology, Pt.1, P.4
43 IBID., P.3
44 Innis, H.A., History of the Canadian Pacific Railway, P.49
45 Dots and Dashes, July-August-September 1974, P.4
46 Daily Free Press, Burlington, Vermont, P.1
47 Canadian Pacific Communication Department, Chronology, Pt.1, P.7
48 MacKenzie, R.J., The Railway Station, P.303
49 Innis, H.A., History of the Canadian Pacific Railway, P.95
50 Canadian Pacific Communication Department, Chronology, Pt.1, P.8
51 IBID., P.10
52 Canadian Pacific Facts and Figures (1937), P.154-5
53 IBID., P.158
54 Dots and Dashes, October-November-December 1973, P.2

Further, Roland Poliquin was in another unique position on Sunday, September 1939 at 5 AM. Poliquin was the operator who transmitted the declaration and entry of Canada into World War Two. The last sentence keyed was: "This country is in a state of war with the German Reich."

CHAPTER THREE - TELEGRAPHIC INVENTIONS

1 Dots and Dashes, September-October 1968, P.3
2 Digby, E., Letter to the Editor, P.112
3 Phillips, L., Yonder Comes the Train, P.346
4 Dots and Dashes, September-October 1968, P.3
5 Dots and Dashes, July-August-September 1973, P.5
6 Railroad Magazine, P.44 (no issue/year)

CHAPTER FOUR - TOOLS OF THE TELEGRAPH

1 Bunnell, J.H., Student's Manual for the Practical Instruction of Learner's of Telegraphy, P.5
2 IBID., P.5
3 Harkness, W.E., The Telephone for Train Dispatching, P.1-2
4 Canadian Pacific Communication Department, Doc. #277, Pt.2, P.1
5 Harkness, W.E., op. cit., P.20-21
6 Canadian Pacific Communication Department, op. cit., P.3
7 Daily Free Press, Burlington, Vermont, P.1
8 Lauckner, Mark, Canadian Railway Telegraph Insulators
9 Dots and Dashes, April-May-June 1977, P.9
10 Warner, D.G., The First Fifty Years of the Canadian Telegraph, P.9
11 Taber, T.T., North to Hudson's Bay, P.28
12 Warner, D.G., op. cit., P.19
13 Horner, G.W., Personal Recollection

CHAPTER FIVE - CIRCUITS

1 Stevens, G.R., A History of the Canadian National Railways, P.331
2 Canadian Pacific Facts and Figures (1937), P.153

CHAPTER SIX - TELEGRAPH PEDAGOGY

1 Compiled, various sources, Burnet, R.G.
2 Compiled, various sources, Burnet, R.G.
3 Compiled, various sources, Burnet, R.G.
4 Stevens, G.R., History of the Canadian National Railways, P.157
5 Dots and Dashes, July-August-September 1987, P.5
6 D'Agapegoff, A., Codes and Ciphers, P.76
7 IBID., P.78
8 Angus, Fred, Canadian Rail, Issue #437, Page 192

CHAPTER SEVEN - TELEGRAMS, SERVICES, FORMS and EMPLOYEES

1 Grand Trunk Railway Rulebook 1911, compiled R.G. Burnet
2 CNR Employee Handbook, P.1
3 IBID., P.1
4 CNR Dots and Dashes, April 1942, P.1
5 Woods, S.E., Cinders & Saltwater, P.202
6 CNR Employee Handbook, P.168
7 Grand Trunk Railway Rulebook 1911, compiled R.G. Burnet
8 CNR Employee Handbook, P.168
9 Horner, G.W., Personal Recollection
10 CNR Employee Handbook, P.149
11 Toronto Star Staff, June 3, 1905, P.1
12 Canadian Pacific Facts and Figures (1937), P.154
13 IBID., P.155
14 CNR Dots and Dashes, October 1942, P.2
15 IBID., P.2
16 CNR Employee Handbook, P.283
17 Innis, H.A., History of the Canadian Pacific Railway, P.249
18 IBID., P.243
19 CNR Employee Handbook, P.III Forward

20 Canadian Pacific Facts and Figures (1937), P.110
21 Dots and Dashes, October-November-December 1989, P.11
22 Warner, D.G., The First Fifty Years of the Canadian Telegraph, P.9
23 Phillips, L., Yonder Comes the Train, P.350
24 MacKenzie, J., The Railway Station, P.303
25 Dots and Dashes, July-August-September 1975, P.9
26 Woods, S.E., Cinders & Saltwater, P.178
27 Railway and Shipping World, March 1899, P.93
28 IBID., P.93
29 IBID., P.95
30 Canadian Pacific Facts and Figures (1937), P.107

BIBLIOGRAPHY

Angus, F., The Canadian Railway Troops in World War One, Canadian Rail, Issue #437, November-December 1993

Batten, J., Canada Moves Westward 1880-1890, Natural Science of Canada Limited, Toronto, 1977

Berry, W.B., From Canoe to Airplane, Railroad Magazine, Pages 46 to 55, undated

British North American Act 1867, Government Documents and Press, Ottawa, 1867 (reprint 1967)

Bunnell, J.H., Student's Manual for the Practical Instruction of Learner's of Telegraphy, Sonneborn Press, New York, 1884
(Reprint Edition of this excellent manual available from L.A. Bailey, Clearwater, Florida)

Burnet, M.J., (nee McCalpin) various personal telegraphic and railway experiences

Burnet, R.G., The Railway Telegraph and Telephone, Canadian Rail #425, 1991

Canada and Its Provinces, Volume 15, Edmonton University Press, Edmonton, 1913

Careless, J.M.S., Canada: A Story of Challenge, MacMillian of Canada, T.H. Best Printing Co., Toronto, 1963

Canada Grand Trunk Telegraph Company, Minutes 1854, Archives of Ontario, Box 8 (g), Pamphlet: Rail and Navigation, Toronto, 1854

Canadian National Railways Employee Handbook, Canadian National Railways, Montreal, December 1956

Canadian National Railways Telegraph Company, CNR Corporate History, Montreal, 1956

Canadian Pacific Communication Department, Chronology Part I, Pages 1 to 12, Montreal, undated

Canadian Pacific Communication Department, Part 2 #277, Pages 1 to 5, Montreal, undated

Canadian Pacific Railway, Canadian Pacific Facts and Figures (1937), CPR Foundation Library, Gazette Printing Co. Ltd., 1937

Copeland, George, many personal railway experiences on the CPR

Coles Canadiana Collection, The Railways of Canada 1870 to 1871, Coles Publishing Company, Toronto, 1970 reprint

Cote, J.G., File Cards, Canadian National Railways DeChief Library, Montreal, undated

D'Agapegoff, A., Codes and Ciphers, Pages 70 to 79, undated

Daily Free Press, Burlington Vermont, Canadian Pacific Railway, September 13, 1886

Dennison, M., and Bell, C., Canadian National Railways Magazine, Vol. XVI, No. 6, June 1930

Digby, E., Letter to the Editor, Railroad Magazine, Page 112, July 1937

Dots and Dashes, Vol. 11, No. 11, Canadian National Telegraphs, Montreal, April 1942

Dots and Dashes, Vol. 12, No. 5, Canadian National Telegraphs, Montreal, October 1942

Dots and Dashes, Vol. 12, No. 8, Canadian National Telegraphs, Montreal, March 1943

Dots and Dashes, Vol. 12, No. 9, Canadian National Telegraphs, Montreal, May 1943

Dots and Dashes, Morse Telegraph Club Newspaper, Fresno, California, Issues January 1968 to 1992

Easson, R.F., The Telegraph in Canada, Canadian National Railways DeChief Library, Montreal, undated

Glazebrook, G.P. de T., A History of Transportation in Canada, Volumes 1 and 2, McClelland and Stewart Ltd., 1964

Green, E., Canada's First Electric Telegraph, Ontario Historical Papers and Records, Volume XXIV, 1927

Harkness, W.E., The Telephone for Train Dispatching, Feb. 12, 1909

Hathaway, R.H., Forward to The Telegraph in Canada, CNR DeChief Library, Montreal, undated

Herstein, H.H., Hughes, L.J., Kirbyson, R.C., Challenge and Survival: The History of Canada, Prentice Hall of Canada, 1970

Horner, George, courtesy many CNR materials, personal telegraphic and railway experiences

Innis, H.A., A History of the Canadian Pacific Railway, University of Toronto Press, 1971

Kennedy, E.M., Train Wires to the Horizon, Railroad Magazine, Pages 8 to 38, undated

Lacombe, D.V., Celebrates Telegraph Centenary, CNR Magazine, Montreal, January 1950

Lauckner, M., Canadian Railway Telegraph Insulators, Lauckner, 1996

Lavallee, O., Van Horne's Road, Railfare Enterprises Ltd., Don Mills, 1974

MacKay, D., The Asian Dream, Douglas and McIntrye, British Columbia and Toronto, 1986

——, The Peoples' Railway - A History of Canadian National Railways, Douglas and McIntrye, British Columbia and Toronto, 1992

MacKirdy, K.A., Moir, J.S., Zoltvany, Y.F., Changing Perspectives in Canadian History, J.M. Dent and Sons (Canada) Limited, Don Mills, 1971

McDougall, J.L., Canadian Pacific - A Brief History, McGill University Press, Montreal, 1968

Mika, N., Railways of Canada - A Pictorial History, McGraw-Hill Ryerson, Toronto, 1983

Murray, J., A Story of the Telegraph, J. Lovell & Son, Ltd., Montreal, 1905

Newman, P.C., Ceasars of the Wilderness, Penguin Books Canada, Markham, 1987

Northern Electric Company, Composite Telephone and Telegraph Systems for Railway Service, Document No. 3853, undated

Phillips, L., Yonder Comes the Train, Galahad Books, N.Y., 1986

Phillips, R.A.J., Canadian Railways, McGraw-Hill Ryerson Company of Canada, Toronto, 1968

Railway and Shipping World, Pages 93 to 95, March 1899

Richards, J., and MacKenzie, J.M., The Railway Station - A Social History, Oxford University Press, 1988

Shaffner, T.P., The Telegraph Manual - A Complete History, Pudney and Russell, New York, 1859

Statutes of the Dominion of Canada, Railway Legislation 1867 to 1915, Dominion Government Press, 1915

Stevens, G.R., A History of the Canadian National Railways, MacMillian Co., N.Y., 1973

Taber, T.T., North to Hudson's Bay, Railroad Magazine, Pages 14 to 34, undated

Thornton, J.J., Keeping Pace with Prosperity, Pages 15-16, Canadian National Railways Magazine, Montreal, August 1929

Toronto Star Staff, Insanity on the Increase, Toronto Star, June 3, 1905

Warner, D.G., The First Fifty Years of the Canadian Telegraph, University of Toronto Press, April 1975

Woods, S.E., Cinders & Saltwater - The Story of Atlantic Canada's Railways, Numbus Publishing, Halifax, 1992

GENERAL CHAPTER INDEX